ROUTEMASTER

First published 1992

ISBN 185414 142 2

Published by Capital Transport Publishing
38 Long Elmes, Harrow Weald, Middlesex

Printed by The KPC Group, Ashford, Kent

Overleaf RM 1101 in Hackney Road in 1975. *Capital Transport*

Above RM 233 at Enfield in April 1974. *Capital Transport*

Opposite RMC 4 at Hatfield in May 1972. *Colin Brown*

Front cover RCL 2222 fresh from overhaul into red livery on its first day in service on the 149. *Geoff Rixon*

Back cover RML 2439 at Elstree newly repainted into NBC corporate livery.

ROUTEMASTER

KEN BLACKER

VOLUME TWO
1970-1989

Capital Transport

Top BEA Routemasters at the West London Air Terminal, Gloucester Road, in March 1971. *Tony Wilson Collection*

Above Hanimex all-over advert liveried RML 2280 at Pimlico in April 1973. *Colin Brown*

ROUTEMASTER

CHAPTER ONE
THE GLC ERA   10

CHAPTER TWO
LONDON COUNTRY   34

CHAPTER THREE
THE FRM   48

CHAPTER FOUR
ALL-OVER ADVERTISEMENTS   50

CHAPTER FIVE
THE XRM – A DREAM UNFULFILLED   56

CHAPTER SIX
NORTHERN GENERAL DEVELOPMENTS   62

CHAPTER SEVEN
AIRWAYS ROUTEMASTERS   66

CHAPTER EIGHT
PRESERVING THE PROTOTYPES   68

CHAPTER NINE
THE QUEEN'S SILVER JUBILEE   70

CHAPTER TEN
SHILLIBEER, SHOPS AND CELEBRATIONS   74

CHAPTER ELEVEN
LONDON'S SECONDHAND ROUTEMASTERS   80

CHAPTER TWELVE
ALDENHAM WORKS   92

CHAPTER THIRTEEN
LONDON BUSES LTD   98

CHAPTER FOURTEEN
LONDON SIGHTSEEING   116

CHAPTER FIFTEEN
LIFE AFTER LONDON   120

AUTHOR'S NOTE

As I stated in the author's note to the first volume, the history of the Routemaster is an ongoing one which hopefully will not be concluded for many years to come. The first volume ceased at the end of 1969; the story now takes up two strands, one of which continues to follow Routemaster developments on London Transport itself whilst the other reviews events on NBC-controlled London Country which took over part of the fleet in 1970.

In a volume of this nature there are limitations on the amount of detail that can be included. Where technical matters are concerned I have endeavoured to cover all major points without going into such a wealth of small detail as to make the narrative uninteresting to all except a small minority of readers. For space reasons it has not been possible to list the routeings or termini of services referred to, or every intermediate alteration or withdrawal; this information can be obtained elsewhere. Nor has it been possible to document within the text all of the many weekend variations from the standard Monday to Friday patterns of operation. Wherever withdrawal dates are quoted these relate, unless otherwise stated, to the first date the vehicle or service referred to is non-operational, it having worked for the last time on the previous day.

Every effort has been made to make this history as accurate as possible but it is an unfortunate fact that, no matter how careful the author may be or how assiduously the finished draft is checked by others, a few mistakes are likely to creep in undetected. A work such as this consists essentially of thousands of facts strung together in what, hopefully, forms a readable story, and it is almost impossible to avoid missing out or misinterpreting the odd one. A particular difficulty arises when even information from official sources proves contradictory, as it sometimes does.

I am indebted to many people who have assisted in one way or another in the preparation of this volume, generally by providing information or by checking the text. I am especially grateful to Lawrie Bowles, Colin Curtis, Ken Glazier, Jim Jordan, Peter Nichols, Alan Nightingale, Bob Pennyfather, Douglas Scott, Colin Stannard, David Stewart, Reg Westgate and Bob Williamson, but particularly to the late Ron Lunn who was my colleague over many years in collating London bus history and whose notes I have referred to frequently when writing the text. My thanks, too, to the many photographers whose work I have used; it is their photographs which bring the story to life.

Lowestoft, January 1992   KEN BLACKER

Opposite Upper RMC 1516 working from Hatfield and advertising the Welwyn Department Store in 1974. *Capital Transport*

Opposite Lower Emergency roadworks at Petersham form the background for this May 1979 view of RM 2130. *Colin Brown*

Upper Right 'Parcel' liveried RM 534, celebrating the wedding of Prince Charles and Lady Diana Spencer, photographed in Park Lane in 1981. *Ramon Hefford*

Right RM 17 in pre-war style livery to commemorate fifty years of London Transport passes Knightsbridge Barracks in 1983. *Geoff Rixon*

INTRODUCTION

Without doubt the Routemaster is the most famous bus ever to run on the streets of London. It has been an active participant on the London scene for more than three decades and, thanks to its longevity, has now become a legend in its own lifetime. In the first volume we traced the story of the Routemaster from its inception soon after the second world war, recording its development through four prototypes, the initial entry into service of the standardised production run in 1959, and the cessation of manufacture in 1968. This volume picks up the threads of the story at the beginning of 1970, a year which saw the de-nationalisation of London Transport and the hiving off of the country part of its once great empire to a new organisation, London Country Bus Services.

The nineteen-sixties had been a period of contraction for the bus industry, and the closing years of the decade found operators everywhere grasping at driver-only operation as a means of diminishing their financial woes. Even the busiest urban routes were not exempt and London Transport, whose commitment to abolishing conductors had once appeared less intense than elsewhere, wholeheartedly flung itself into a major urban conversion scheme with its Reshaping programme from 1968. In such a seemingly hostile environment, the Routemaster's front engine, rear entrance configuration seemed to spell imminent doom, a fate which was seemingly confirmed by London Transport's expectation of ending all crew operation by the end of 1978.

Events have shown all too clearly that this gloomy forecast was destined not to be fulfilled. Routemasters saw the nineteen-seventies through, and the nineteen-eighties too, and even though many met their end from 1982 onwards, a substantial number continued serving central London routes as the decade came to a close. Furthermore, their operating sphere had widened to embrace previously undreamed of horizons as many provincial towns began echoing to the familiar and mellow strains of the Routemaster for the first time.

The fact that the Routemaster has lasted so long is in no small measure due to the totally inferior designs which came afterwards and which cost London Transport very dearly in premature write-offs, destroying public confidence in the process. Many of these later designs suffered from insufficient development funding and prototype testing, inadequate specification and, perhaps most important of all, inadequate consultation with those who would have to operate and maintain them. While London Transport was wrestling with the deficiencies of its newer rolling stock the wheel turned its full circle; thirty years after it first flourished the Routemaster was once again in demand.

The Routemaster is world famous as the archetypal London bus; it is still with us and likely to be around for several more years. In a story which is still ongoing it has been necessary to establish a cut-off point for Volume 2, and for this purpose 1st April 1989 acts as the end of our narrative for the time being. This was the date on which unified operation of London's bus services was finally devolved to a dozen subsidiaries and when the old structure which had been in force since 1933, through alternating periods of glory and difficulty, was brought to an end. Ahead lay an uncertain future of possible service deregulation and company privatisation, the exact form which these would take, if any, being dependent upon the prevailing political climate. Long term planning, such as may be deemed vital for the future well-being of any industry, cannot be made in such circumstances, and it is certain that in the absence of stability the inspired commitment and financial dedication such as produced the Routemaster will not again be forthcoming. The Routemaster cannot, of course, go on for ever but it is by no means beyond the realms of possibility that at least a few of the class may still remain active into the next century. The fact remains that, for very intensive in-town work where operating economy, threat of competition, or political expediency dictate that a two man crew is still desirable, the Routemaster meets an ideal which has not been bettered.

Facing Page

RMA 19 with Bus Engineering Ltd at Chiswick. *John Laker*

RMs 143 and 84 during their first season on the sightseeing tour. *Ramon Hefford*

RMC 1456 on commuter service 15B at Mansion House. *Stephen Madden*

RM 44 in June 1970, when it was a recent recipient of the plain block capital London Transport fleetnames introduced at the beginning of the year and when Red Bus Rovers and a trip on the sightseeing tour each cost seven shillings (35p). Twenty years later, despite keen competition on the latter, the prices were £2.30 for an all-zones bus pass and £8 for the tour. *Capital Transport*

CHAPTER ONE
THE GLC ERA

The transfer of London Transport from State to municipal control on 1st January 1970 was widely perceived as holding out the best hope for a revival of the undertaking's flagging fortunes. As things transpired, the GLC reign produced a curious and unstable mixture of ups and downs. On the positive side there was innovation in vehicles, fares and services, heavy capital investment, and a determined effort to stem the chronic staffing shortage by paying realistic wages, introducing improved facilities and mortgage schemes for staff, and by employing women as bus drivers (after overcoming Trade Union opposition in 1973). Countering all this were abrupt policy changes and constant political interference from County Hall such as led to the resignation, at the end of 1974, of Sir Richard Way, the most able and well respected person to have held the chairmanship of London Transport since the formidable Lord Ashfield. The 1970s were years of increasing assaults on staff and vandalism of vehicles whilst, in the middle of the decade, bomb alerts became a regular feature of London life. This was the scene prevailing as the Routemaster family entered its second decade which was also intended to be its last. London Transport's annual report for 1970 presupposed the demise of the Routemaster in stating that "By the end of this decade, every London Transport bus will be operated by one man and the total bus operating staff reduced by 10,000". The end of 1978 was in fact the unofficial target for the end of crew operation.

The final eighteen months of the old regime had witnessed a transformation with hundreds of new Merlin single deckers flooding into service, spreading one-man operation far and wide. The final members of the MB family were delivered in 1969 but the shorter and less powerful Swift was coming on stream, the first SMs entering service at Catford on 24th January 1970. With double deck one-man operation also starting from 2nd January 1971, introducing the first of a large contingent of Fleetlines, the future for crew operated buses certainly looked gloomy, and though large numbers of RTs still had to be disposed of, London Transport looked to be well set on its course for eliminating the Routemaster before the decade was over.

January 1970 saw a slightly revised image for Routemasters with the introduction, as a standard feature for the class, of a plain block capital fleet name without underlining, as already carried for some time by FRM 1. Stated as being only for more modern buses in the fleet, the new style, authorised by London Transport's internal design committee of so-called experts, was seen by many as merely one of a series of retrograde design policy decisions gradually distancing London Transport from its high standards of earlier days. In the context of the many problems confronting the organisation, not least of which was the loss of almost ten per cent of scheduled mileage through driver and conductor shortage, such tinkering with appearance seemed irrelevant.

The first service change of the new era occurred on 24th January but, before this, RMs had succeeded in ousting the hapless XA class Atlanteans from central London service. Withdrawal for conversion to driver-only operation on less demanding suburban runs had commenced during the autumn of 1969 and very few survived into 1970, the last being withdrawn from route 67 at Stamford Hill and routes 34B and 76 at Tottenham on 21st January. The 24th January changes saw the withdrawal of a number of RM routes; the 34B was a casualty along with routes 127 (worked by Enfield and Edmonton), 239 (Chalk Farm) and 283 (Norbiton). The final batch of Merlins was put into service on new Red Arrow service 513 causing further RM surpluses from routes 7 and 13 which were shortened to make way for the Red Arrows whilst route 134 lost its in-town end between Warren Street and Pimlico in deference to the new Victoria Line tube. Merton garage obtained its first RMs for a new allocation on route 131 and Turnham Green gained a share on route 117 which also resulted in an RM presence on night route N97. In south-east London, Plumstead lost its RMs along with its share of route 53. Routes 298, now with altered termini at both ends and a joint Palmers Green and Wood Green allocation, and 298A worked by Palmers Green alone, were converted from RT to RM but many more RMs were temporarily delicensed, to reappear on 7th February at Norwood, Croydon and Chalk Farm as replacements for RTs on route 68.

On 18th April 1970, Twickenham garage closed, the first closure for some years, without ever having enjoyed an RM allocation. On the same day, RMs displaced from Hounslow's 81 group by SMs (except for the 81C which ceased) enabled Middle Row and Wandsworth to receive RMs for route 28 and, at Wandsworth, route N88. The same date saw the introduction of the flat-fare, XA worked C-route network in Croydon resulting in many of Croydon's RMLs going into temporary storage whilst new work was found for them. Between 1st and 8th June, 14 surplus RMLs joined no fewer than 27 RTs for an ambitious series of road layout trials at the Transport & Road Research Laboratory's premises at Crowthorne, Berkshire.

13th June 1970 saw the restoration of full RML operation with the reintroduction of the class at Stamford Hill and a first-time allocation at Wood Green, both for route 243. Numerous service changes in north-west London, many resulting from an influx of new Swifts, included withdrawal of route 8B in favour of semi-express MBS route 616, but this was more than compensated for by new routes 26 (Finchley and Cricklewood) and 32 (Cricklewood), both of which were RM worked. In pursuance of the policy of removing RTs from central London operation, routes 27 (Holloway and Turnham Green garages) and 159 (Camberwell and Streatham) both received RMs. On 18th July, it was the turn of east London to participate in major service revisions and, though these did not bear greatly on the RM fleet, the reallocation to Upton Park of Barking's workings on route 23 resulted in the loss of all RMs and the subsequent reversion of route N95 to RT operation. These were the last changes to affect the Routemaster fleet during 1970 but a major portent for the future occurred on 30th September with the delivery of the first DMS. These box-like vehicles, with squared-off looks inspired by Manchester Corporation's contemporary 'Mancunian' design but lacking the imaginative livery to enhance the new shape, were launched as the new 'Londoner' bus but, being generally perceived as ugly and uncomfortable, the efforts at endowing them with a title to place them on a level with the Routemaster flopped hopelessly and were quickly dropped. The DMS, with its capability for one-man operation, was intended to have a major impact on the Routemaster family. The impact, when it came, was totally the opposite of what had been expected. Instead of dooming the Routemaster to extinction within a few years, the troubles which quickly beset the Fleetline were such as to contribute to a stay of execution beyond the critical decade, right through the next and into the 1990s.

A design modification introduced starting in December 1970 was the omission of the leather ends from crosswise seat cushions on economy grounds. This enabled cushions to be interchangeable between offside and nearside, subject to the fitment of a common locating block.

Although the legal parting of the ways between the old central and country bus fleets occurred overnight, at ground level it took some time to become fully effective, so close were the old established bonds. For a while, when breakdowns occurred on Green Line, help continued to be sought for the provision of replacements as witnessed by Highgate's RM 428 helping out on the 715 at Shepherds Bush in July 1971. The 'On Hire to London Country Bus Services Ltd' sticker in the front window, not readable in the photo, was now a necessity because of the different ownerships and inter-company invoices were now required to cover the cost. J.H. Blake

An early casualty of the 1970s was route 81C which was withdrawn in April 1970 through lack of patronage. Hounslow's RM 1023 is seen heading to take up service on this operation which lies on the very western fringe of the red bus area. E. Shirras

1971 was only two days old when the first DMSs entered service. The two routes selected to herald this new dawn for London's transport were both Routemaster operated; the 95 worked by Brixton garage had once been a major tram route and, more recently, the last home of the RTW class, whilst Shepherds Bush's 220 was a former trolleybus service of equal importance. Almost immediately, problems manifested themselves on the new Fleetlines, bringing much reduced reliability compared with the displaced RMs, the majority of which were earmarked for route 38 (Clapton and Leyton garages) to replace RTs from 16th January, while Camberwell's route 59A was also due for early conversion to RM. At Brixton, a by-product of the loss of its RMs was the reversion to RT of night route N87. The third Routemaster to DMS conversion occurred on 16th January and actually restored Fleetline operation to Highgate's route 271, members of the XF class having worked it experimentally alongside Atlanteans although with conductors and not in the new driver-only mode. RMLs rendered surplus from route 271 moved only a short distance to Muswell Hill and continued serving much of their former territory on route 43 which had major sections in common with route 271.

13th March was the next date in 1971 on which DMSs replaced RMs, the route in question being Cricklewood's 32 which had commenced only in the previous June. A fortnight later, a shortened route 196 became RM worked from Camberwell instead of RTs, the northern end and its Highgate allocation being taken over by SMSs under the revived route number 239. On 17th April, the onward march of DMSs ended the RM regime on route 5 (Poplar and West Ham garages) whilst West Ham also received SMSs to oust RMs from route 238. Next in a year of many changes came the conversion of Leyton's route 48 from RT to RML on 15th May using ex-Southall buses rendered surplus by the withdrawal of route 207A whilst, on the same date, a shortened route 49 received RMs in place of RTs at Streatham and Merton garages. A feature new to RMs in May was the use of white, DMS-style open roundels in substitution for the plain gold fleet name introduced less than eighteen months earlier. One hundred RMs received the new treatment as they passed through the Aldenham paint shop, after which the former style was resumed. On the nearside, the roundel was applied in the customary midway position on the lower panels but, on the offside, it was positioned at the rear by the staircase leaving the lower panels looking rather bare.

The second half of 1971 saw new double and single deckers flooding in, the next move affecting RMs being on Catford's route 124 which surplus members of the class took over on 24th July. 4th September saw the second and final garage closure of the decade when Holloway ceased to function, its operations passing to the nearby ex-trolleybus Highgate premises which, though retaining code letters HT, was renamed Holloway, resuming the title which it had held up to the merging of the Central Bus and Tram & Trolleybus departments in 1950. On the same date, a reallocation on route 243 resulted in the loss by Wood Green of all its RMLs to Tottenham whilst the

conversion to DMS of Fulwell's route 267 provided RMs for route 90B. On 30th October, the RMs on West Ham's route 162 succumbed to SMS single deckers, making it possible to convert Croydon's route 190 from RT to RM on 20th November. The final moves of 1971, on 4th December, ended the operation of RMs on Thornton Heath's route 64 and RMLs on Stamford Hill's route 67, both of which fell victim to the onward march of the DMS. However, Stamford Hill gained a new RM route, 97. By the end of the year, most Routemasters in the fleet, in common with all other types, had lost their rear wheel discs under an edict dated 18th November that these, and their retaining brackets, should be removed immediately. Two garages which responded more slowly than most to the edict, Wood Green and Palmers Green, finally complied fully early in 1972.

Top RMLs found their way onto Leyton's route 48 in May 1971, having been made surplus with the demise of the 207A. Route 48 was the first direct conversion from RT to RML apart from the temporary 1965 allocation to Tottenham's routes 34B and 76 as part of the Routemaster/Atlantean comparative trials. RML 2713 is about to set off from London Bridge station in this August 1973 view. Tom Maddocks

Above **RM 291** was one of Holloway's fleet which moved across to Highgate (renamed Holloway) on closure of the smaller property on 4th September 1971 and had already acquired its HT garage code when photographed six days later. Route 27 had itself only been RM operated for a little over a year at this time. Alan B. Cross

A common sight for only a relatively short time were the 100 Routemasters decked out with open roundels in May and June 1971. Peckham's RM 181 and New Cross's RM 168 show the offside and nearside positioning of these respectively.
Tom Maddocks/
Capital Transport

Conversion to DMS of route 90B in January 1973 enabled a start to be made on displacing RTs from route 12, and not before time. It was strange that many comparatively quiet suburban routes had received modern rolling stock ahead of such a major cross-London one. RM 1017 was one of the vehicles transferred from Fulwell to Walworth on this occasion. RM 5 (below) was one of the vehicles already at Peckham when its share of the route was converted. Colin Stannard

Outwardly, things remained much as normal at the start of 1972, with the onslaught on the elderly but still numerous RT family continuing apace, either directly through the influx of new vehicles or indirectly by the reallocation of Routemasters, as occurred on 8th January when the trunk route 25 worked by Bow and West Ham was turned over to RMs. Some were relicensed after having been made surplus by earlier changes but others came from Catford whose route 124 became DMS operated less than six months after being taken over by RMs. Also on 8th January, Peckham lost its RMLs to Stockwell as a result of a reallocation of work on route 37. A new RM route introduced on this date was the 46, a Chalk Farm operated breakaway from the northern end of the lengthy and circuitous route 45 whose shortened remainder was now worked solely by Walworth. On 5th February, the final batch of Swifts entered service, sadly severing, after sixty years, the strong bond which had existed between London Transport and AEC to the great benefit of both. However, the Routemasters were unaffected by this and, indeed, were out of the limelight for several months. The problems over procurement of spare parts, which were to hit the whole industry so badly, were just beginning to surface and, compounded with a falling level of attention towards day-to-day maintenance because of the need to spend so much time on the new rear engined types, the net result was a growing shortage of serviceable Routemasters. London Transport was now experiencing NBA (No Bus Available) problems which, until quite recently, would have been unthinkable and this prompted a reassessment of the role of the RT class and even resulted in the purchase of 34 such buses from London Country in September 1972 as a short term expedient for overcoming the immediate problem. Later in the year, in a realistic appraisal of its long term strategy, the Executive came to the conclusion that omo conversion of busy routes in central London could not be undertaken for the time being at least.

11th March 1972 saw the withdrawal of Camberwell-worked route 59A following which, on 17th June, RMs were removed from three more services. Riverside's route 255 was withdrawn entirely while the 46 at Chalk Farm and 295 at Shepherds Bush were handed over to DMSs. Chalk Farm's RMs stayed on to displace RTs from route 31 for which Battersea also received RMs. The time was now approaching for an RML overhaul float to be established and it was decided to obtain the required vehicles from route 262 which did not need the larger capacity and could easily be worked by 64-seaters. On 1st July, West Ham's share was dealt with, many of the incomers being ex-float RMs including some bearing identities which had not been seen for almost a decade. Leyton's workings on route 262 became RM on 15th July using buses from Finchley thrown up by the conversion to DMS of route 253. August 12th saw the conversion of route 106 (delayed from February), RMs at Tottenham and Hackney passing to Holloway and Battersea in place of RTs on route 19. Finally for 1972, on 28th October, West Ham's route 278 was taken over by DMSs allowing RMs to pass to Upton Park for route 101. As the year ended, a minor livery change introduced white as the standard colour for the central band on Routemasters, giving somewhat greater impact than the rather drab grey of the past few years.

Next on the list for conversion from RT to RM was route 12, a four-garage allocation requiring more than sixty buses which was tackled in stages occupying almost the whole first half of 1973. The arrival of DMSs on Fulwell's route 90B provided RMs for Walworth's share of route 12 on 6th January. Thereafter came a two-month lull up to 10th March when Elmers End was converted using RMs displaced from Edmonton and Tottenham by one-manning route 259 and from Putney's route 85A which was dealt with likewise. DMS conversion of route 221 on 24th March provided RMs from Wood Green and Finchley for Peckham, following which came another lapse of almost two months before DMSs ousted RMs from Norbiton's route 131, enabling Shepherds Bush to be converted for the completion of route 12. On 12th June, West Ham's route 241 was converted to DMS on its Docklands end whilst new RM route 230 was inaugurated to replace the long northern leg from Stratford to Manor House. At the same time, in an endeavour to minimise the effect of traffic delays between Walthamstow and Chingford, routes 69 and 262 were replaced over their common northern end by localised route 269 which was also RM worked. A further DMS conversion on 16th June, that of Hornchurch's RML worked route 165 was of particular significance in bringing into existence London Transport's first all-omo garage, starting a process which was to continue unabated for many years to come as the conductor's role steadily diminished.

However, there were to be no more driver-only conversions in 1973 to impact on the Routemaster fleet. In the face of an accelerating shortage of serviceable buses, with nearly two hundred off the road by early August awaiting mechanical attention and spare parts, even the disposal of RTs had reduced to a trickle. In June, the sale of Merlins commenced with the original Strachans bodied batch, and this was followed in August by the announcement that the whole of the large Merlin fleet was to be withdrawn; 938 new buses, including 843 double deckers, would be ordered to replace them along with the remaining RTs and RFs. This marked a significant swing back to double deckers and was tacit admission that the mass move into single deckers which had brought about the end of Routemaster production in 1968 had been ill advised. It was also further confirmation of the retention of Routemasters, and a significant amount of crew operation, for some years to come while new buses were required to replace unsuitable omo classes. As it was, the current omo programme was proving hard to implement because of the growing shortage of drivers and conductors which, having reached a crisis level of about 4,500, was making very difficult their release for training without further reducing the service on the road. A nine-point plan of improvements announced by the GLC, including massive wage increases, promised well for the future but, meanwhile, morale throughout the organisation was at rock bottom as witnessed by a ten-week dispute at the Chiswick engine overhaul shop in protest at the low level of bonus payments, which only exacerbated the already serious vehicle shortage. Towards the end of the year, it was decided that some of the new Fleetlines coming on stream would be put to work as crew operated vehicles as a means of getting them operational as quickly as possible to ease the worsening vehicle position. Until the DM version of the class was available, modified DMSs were used, commencing on 15th December with routes 16 and 134. The RMs thus displaced were reallocated to the 77

New RM route 230 brought this route number to highbridge double deckers for the first time when it was introduced in June 1973. During that summer, RM 313 is seen at Manor House. J.G.S. Smith

An interesting working of long standing was the operation by Merton garage of a few morning peak journeys on route 164 using buses from route 77. When the 77 received RMs so did the 164 journeys, even though the main Sutton allocation was destined to retain RTs for another three years. RM 1850 is seen at Morden station. Alan B. Cross

group, Cricklewood's from route 16 to Merton, and Muswell Hill's and Potters Bar's from route 134 to Stockwell. The Merton vehicles also worked scheduled morning peak journeys on route 164, the major part of which continued to be RT worked from Sutton. RTs were now a common sight deputising for Routemasters at those garages which had both, but the surprise arrival late in 1973 of an RT at Tottenham, later joined by two others, after a lapse of five years, indicated the growing seriousness of the position.

AEC Merlins, on which all hope was pinned in the late nineteen sixties, dominated places such as Walthamstow Central bus station for only a few years whereas RMs have continued to soldier on. In April 1975 RM 271 still looks in good fettle but life for the Merlins is almost over. MBS 456 on the W21 has only a few days left to go whilst MBs 347 and 380 will also be withdrawn before the summer is out. Ken Blacker

A surprise conversion to RM in October 1974 was the lightly used 175A of which a long section ran through open countryside where passenger demand was sparse. However it only required one bus as depicted by North Street's RM 1426 heading for Chipping Ongar with barely a passenger on board. The white roundels and fleet numbers made their appearance in April 1974.

Routemasters barely featured in the news during 1974 apart from when, on 2nd February, crew-worked DMSs took over route 149 pending the arrival of DMs which did not commence until September with the first intake concentrated on route 16. It was originally intended that the garages operating on route 149, Edmonton and Stamford Hill, would despatch their RMs to Bromley, Catford and Dalston for route 47 but this was not to be. RTs stayed secure on route 47 while the RMs were distributed far and wide to fill in gaps left by Certificate of Fitness expiries which the overburdened engineers were unable to cope with. A similar fate had earlier befallen the RMs and RMLs released by the 165 and 241 omo conversions back in June 1973 but a backlog of delicensed Routemasters which had stood at almost one hundred by late autumn continued to increase into 1974. The only route to gain an RM allocation during the year was North Street's 175A early in October and this required only one bus. A sign of the deteriorating social order was the allocation, at the crews' request, of DMSs instead of RMs to Cricklewood's N94, from 1st March, for the greater security that a doored bus would bring. Towards the end of the year, hooliganism resulted in Middle Row's 48 RMs being the first to be fitted with assault alarms operated by the driver to give a continuous blast on the horn and flashing lights to deter aggressors and to indicate that help was needed.

1975 saw the start of a programme for fitting two-way emergency radios to all vehicles in the fleet.

What had been termed a "comprehensive design study" into various aspects including bus livery resulted in the adoption, for more modern buses within the fleet, of white instead of gold fleet numbers, and the use of a plain unlettered white roundel instead of a fleet name. As far as the Routemaster family was concerned, these features were first tried out on experimental RM 8 but were duly adopted as standard for all newly-painted buses from April 1974 onwards. The overall effect, which was brash and unsympathetic to the classic lines of the Routemaster, jarred at first but quickly became an accepted feature of the RM scene. Less noticeable alterations were also effected on various buses during the same period, one of which was the sealing of the lower part of the driver's windscreen with angled metal plates to prevent it opening on buses where repair to the winding gear would otherwise be required. From January 1975, the lower saloon ceilings, where they required complete repainting, were done in white rather than yellow, in the style applied from new to later RMLs; it is believed that RM 1373 was the first to be so treated. Between August and October 1975, 46 RMs originally classified as 5/7RM7/5 were converted on overhaul from Simms to CAV electrical equipment and were recoded 2/5RM5/8. Simms was now defunct as a manufacturer and this substitution provided a source of spare units needed urgently for maintaining the many other Simms equipped buses in the fleet. Most buses concerned were in the range RM 1557 to 1744 although RM 721, 732, 746 and 1000 were included. A very visible modification which appeared on many overhauled buses from late 1975 and quickly

became commonplace was the fitting of floor patches made of unpainted aluminium chequer plate to avoid the time consuming task of renewing the original Treadmaster material where it had become worn beyond acceptable safety limits.

In terms of the crisis over vehicle non-availability, 1975 turned out to be the worst year yet, with a daily shortage for service peaking at about 900 buses by September, although this was reduced to about 600 by the end of the year. The spare parts crisis continued unabated, forcing London Transport's buyers to look worldwide to ease the situation, a notable result of which was the placing of a huge order with Federal Mogul of the USA for Routemaster piston rings. New bus deliveries from Leyland were inexcusably late for all customers but London Transport was worst hit of all with a shortfall of no fewer than 400 new buses by late summer. In order to break free from the Leyland stranglehold, 164 Scania powered Metropolitan double deckers had been ordered from MCW in 1974 but delivery of these was not due to commence until the very end of 1975. Despite all this, large numbers of time expired RTs were withdrawn in 1975, there being no capacity to recertify them, and RMs were largely instrumental in reducing the number of RT garages during the year from 37 to 26. Similarly, the number of Merlin garages dropped from 18 to 10, although this fell far short of the Executive's intention to withdraw all of this class by the end of 1975. Despite the fact that many Routemasters lay around, out of ticket and unlicensed, and in some cases heavily cannibalised, while other still licensed buses were forced into idleness through the want of maybe one or two simple parts, the class still managed to play a very active role in 1975.

A hot day in July 1976 sees RM 295 in Deptford with all its windows and the bonnet flap open. Route 47 and route 1 had been converted from RT to RM at the beginning of 1975. Capital Transport

RML 2302 was among a number of RMLs allocated to Stockwell for route 88 in March 1975 following conversion of the Shepherds Bush allocation to the type in the previous month. P. Picken

26th January 1975 saw a large RM influx into south-east London when route 47 was at last converted at all three of its operating garages (Bromley, Catford and Dalston), with a start also made on route 1 at Catford and the conversion of New Cross' small route 151. The availability of so many RMs was brought about by the placement of new DM crew operated Fleetlines at Muswell Hill for route 43 and at Holloway for routes 17 and 214 (also night routes N92 and N93). The buses deposed from route 43 were RMLs which were transferred to Stonebridge for route 18 where they replaced RMs. On 23rd February, route 45 at Walworth was the next DM recipient, RMs passing to New Cross to complete route 1, and also to Middle Row and Alperton for route 187. The latter move was a surprise, the heavy loss making 187 being far more suitable for conversion to driver only operation than to RM but it was carried out to ease the maintenance position at Alperton by reducing its vehicle types to one during a complex garage rebuilding operation. Later in the year, nine of Alperton's RMs were temporarily transferred to Stonebridge for maintenance and overnight storage, and though still worked by Alperton crews and carrying ON garage letters, they were distinguishable from proper Alperton buses by spots of white paint on the bottom corners of the stencil holders and also carried the SE code painted in black inside the driver's door.

Also on 23rd February, a start was made in converting two additional routes to RML with the arrival of some at Willesden for route 8 and at Shepherds Bush for route 88. The source of these additional RMLs was the ending of the first overhaul cycle and the release into service of the works float. Many months were to elapse before route 8 received any more RMLs but route 88 was completed on 2nd March when Croydon's remaining RML stock from routes 130/B was despatched to Stockwell upon the arrival of new DMs. Stockwell's RMs moved across to Sidcup and New Cross to relieve route 21 of its RTs. The RMLs at Stockwell, though allocated for route 88, immediately became commonplace on route 2B marking the start of an unofficial mixing of RMs and RMLs which was to be practised by many garages in subsequent years.

A summer respite lay ahead before the one-manning on 19th September of Turnham Green's night route N97 and the commencement, three days later, of hired bus operation alongside RMs on Croydon's route 190, using Leyland PD3s belonging to Southend Corporation. This move, the first since the crisis days of the late 1940s to ease chronic bus shortage by hiring buses from other operators, met with stiff union opposition which precluded the hiring of further buses from other sources. T&GWU hostility also prevented the peak period hiring of coaches and drivers on all but two of 34 planned routes. Despite its protestations over the appalling service brought about by vehicle non-availability, the Union was unwilling to back such drastic measures to improve the position. Though staffing levels were now much higher thanks to wage increases and other measures, many crews were spending much of their working day sitting in canteens, on full pay, with no buses to drive or

conduct. On 29th September, reduced schedules were introduced on 110 routes to reflect vehicle availability, though few RM routes were included and the class continued to be in short supply with RTs often used as substitutes. 11th October saw the establishment of a Routemaster presence on North Street's route 175 with the entry into service of the RMA class, newly purchased from British Airways and still in orange livery. Unpopular with the staff, these unusual buses proved unreliable at first and suffered many breakdowns, probably as a result of being laid up for about a year prior to purchase, but they improved with time. On 19th October, route 24 was converted to DM, Chalk Farm's RMLs passing to Bow to complete route 8 although they often appeared on route 8A, and also to Muswell Hill whose previous RMLs had departed only nine months earlier, but now intended for route 13. At long last, Norbiton's route 65 was converted to RM, some seven years after it was first mooted and several intermediate false alarms. On Sundays, Kingston garage ran a small, three bus, allocation on route 65 for which it took the unusual step of borrowing RMs each weekend in preference to using its RTs. On the same date, a redistribution of work involving several garages and aimed at placing work where staff recruitment was easiest, resulted in New Cross receiving RMLs for the first time to cover a new allocation on route 37. At around this time, Stockwell's

night duties on routes N68/81/87/88 were converted from Routemasters to DM. The final change in 1975 was on 14th December when a DM influx on to route 29 at Holloway and Wood Green found RMs passing to Streatham for route 118 and to Riverside for route 72.

Below Left Route 175's flirtation with RMAs in 1975 was not a happy one and RTs were often substituted. In this scene RM 95 is seen playing its part even though the correct blinds are clearly not available! David Stuttard

Below Right By 1975 BESI (Bus Electronic Scanning Indicator), the first attempt at establishing a bus recognition system dating from the early nineteen sixties, was drawing to the end of its usefulness. Dalston's RM 1992, fitted out with reflective plates denoting route and running number by binary code, passes one of the pregnant-looking scanners located at Hyde Park Corner. Capital Transport

Bottom Left A small scale imitation of the previous decade's flirtation with roof adverts occurred in October 1975 when Holloway's RM 473 was employed to publicise the film 'Man from Hong Kong'. G.F. Walker

Bottom Right An experiment was carried out during the winter of 1975/6 in which the upper air intake on three Riverside-based RMs was completely blanked off to improve interior heating. One of these, RM 84, is at Trafalgar Square in November 1975; the others were RMs 1442 and 1887. J.H. Blake

The March 1976 takeover by RMs of Walworth's allocation on route 176 meant that, for the first time in almost 29 years, regular daytime cross-London operation by the venerable RT class was no longer scheduled. RM 1336 was one of the original large batch of Leyland-engined vehicles but was by this time AEC powered. In later post-overhaul reincarnations it reverted to Leyland in August 1979 and back again to AEC in May 1983. Capital Transport

1976 remains in history as the year of the revival of the RT class when many of these veterans came to the assistance of the Routemasters in a big way. Starting in December 1975 with the allocation of a single RT to Norbiton to assist on route 65, the following year saw RTs returning to a whole string of garages to which they were no longer officially allocated, in a manner reminiscent of the revival at Tottenham a couple of years earlier. As most garages concerned had no RT blinds, a whole variety of makeshift displays was seen, some well executed with neat masking to accept Routemaster blinds and others more unusual. In addition, garages which still officially held mixed RM and RT allocations frequently held spare RTs to cover RM operations but these, of course, carried correct RT blind displays and were less obvious. Some routes which had never had RTs, such as Finchley's 26, found themselves with an odd one or two, but most noteworthy in this category was Leyton's route 230 which went over completely to RT from 29th November 1976. So too, a little earlier, did North Street's 175 on 4th September when the RMA class ended its short spell of stage carriage service although, in this case, RTs had never really been absent and were always around in varying numbers to help out.

On 31st January 1976, the Walthamstow area localised route 269 was withdrawn, after a fairly brief existence, with the extension back to Chingford of route 69. 28th February saw the end of Hanwell's RMLs on route 207 with the arrival of new DMs, the first of which entered service a day early on the 27th, allowing displaced RMLs to do likewise at Hendon on routes 13 and 113. Also a recipient of RMLs was route 23, where the West Ham allocation marked a revival for the class at this east London location. Some deposed RMs passed to Seven Kings for route 86, breaking the monopoly at this location of RTs, the only buses which the garage had been physically capable of operating until its reconstruction was completed. Seven Kings' night route N98 also became RM operated, but the reluctance of staff to cover it for fear of assaults led to its transfer to Barking garage in September and a resumption of RT operation. Route N82 was scheduled to receive RMLs on 28th February and, on the same date, Croydon's route 190 reverted to RM operation as the Southend Titans left for a stint at Harlow on loan to London Country. A minor RT to RM conversion, also on 28th February, was Upton Park's route 169A but, far more important historically, was the start of the conversion of route 176 through its Willesden allocation. This was the last scheduled RT operation, apart from night services, across central London and it

was completed on 22nd March when Walworth's share of this and peak period offshoot 176A received RMs.

The RMs for Walworth came from Peckham where the new MD class had begun operation on route 36 some four weeks later than originally hoped due to training delays. The new Metropolitans which quickly became universally popular, had riding qualities which equalled or even bettered the Routemasters, acceleration which was the next best thing yet to a trolleybus, saloon heating which was truly effective, and quietened motive power units. Despite initial torque converter troubles, poor fuel consumption, and worrying early signs of underside corrosion, they were the first double deckers to break the mould of mediocrity which had been the norm since the Routemaster went out of production. Unfortunately, even with its very superior turn of speed, the MD proved slower than the Routemaster on crew operation because of its doors, and questions as to the wisdom of placing Fleetlines on such work because of the deterioration in timekeeping which always resulted were also posed in regard to the MDs. The Metropolitans provided proof, if such were needed, that even a superior type of modern bus was no match for the Routemaster on intensive in-town work where fast scheduling was of the essence.

On 28th March, MDs took over from RMs on route 36A, whilst a progressive conversion of route 36B commenced on 13th April but was not complete until September. More DMs took the road on 28th March, this time at Uxbridge to complete route 207, allowing RMLs to pass to Upton Park to complete route 23. Sutton thereupon received its first RMs, replacing RTs on route 93. On 31st March, the first production Routemaster to be built, RM 8, earned its place in history as the last to enter service when it did so at Sidcup on route 21. A shadowy figure for so many years in its role as experimental shop hack, RM 8 was now replaced in this role by RM 1368. All experimental features were removed from RM 8 and the bus was overhauled at Aldenham before entering service, although unlike the normal Aldenham overhaul, it retained its original body. RM 1368 which took its place was unique in having been converted into a single decker. In its original form, its upper deck had been gutted by an arsonist while parked at Tottenham garage late on New Year's Eve 1974 and, after appraising the remains following removal to Aldenham, it was decided in 1975 to convert it to the role which has since earned it fame as the most unusual looking member of the Routemaster family.

The continued influx of MDs at Peckham produced sufficient RMs to convert Bromley's route 119 and New Cross's 192 on 2nd May, and for a start to be made on Brixton's share of route 109 from the 23rd which took through to August to complete. Meanwhile on 12th June, Sidcup's route 51 was dealt with. On 27th September, Peckham's route 63 received MDs enabling RMs to take over route 193 at Seven Kings from 3rd October and to commence displacing RTs from Thornton Heath's share of route 109. October 1976 was the month in which official withdrawal of the AEC Swifts commenced and, in the following month, London Transport applied to the GLC for authority to dispose of all of them between 1977 and 1979. It was also made known that the DMS class, which was still under delivery, would commence withdrawal from 1980. Though 1976 ended with the vehicle availability position still at about the same critical level as a year earlier, there was now the knowledge that, in the face of failure of most of the modern vehicle types, the Routemaster was set for a further reprieve. Speculation was now rife that they might last through to the end of the 1980s.

Top **On 31st March 1976 RM 8 at last entered service on Sidcup's route 21 over seventeen years after it was built, probably establishing some sort of world record in the process. With all experimental features removed it looked the same as any other Routemaster but still retained its original body.** Paul Hulyer

Centre and Right **The strange looking replacement for RM 8 as the guinea pig for experiments was RM 1368, a neat if somewhat top heavy looking conversion to single-deck having been carried out following the gutting of its upper deck by vandals. A welcome embellishment was the use of traditional gold fleetnames and numerals.** Alan B. Cross

The bus industry was now well into the era when local authorities provided sometimes quite substantial sums of money to keep socially desirable but loss making bus services operational and, outside of the GLC's territory, London Transport was as dependent as any other operator on rural service grants. At the request of Essex County Council, a scheme of revisions on 8th January 1977 included the conversion of route 175A from RM to single deck omo using the new BL type Bristol LH and, at the same time it was re-numbered 247B. Next day, New Cross garage became the second to operate MDs as the final members of the batch were received from the manufacturers over the ensuing weeks. RMs were thereby made available from route 53 for Merton's route 155, followed on 16th January by Sutton's 164/A and, over the next few weeks, by Sidcup's 51A. The latter, however, was to be a short-lived RM operation as it disappeared entirely on 21st May when it was absorbed into the 51 and converted to DMS. Meanwhile, back in January, West Ham's N99 officially became RML worked in place of RM whilst, on 26th February, route N82 became DMS instead of RML. A big change-around on 19th March saw the resumption of RM working on route 29 in place of DMs which were removed after an alarming drop in passenger traffic due to the use of doored buses. The DMs were used instead on route 141, Wood Green retaining its own and Holloway's passing to New Cross who often managed to produce an MD to add variety. Route 29 also gained a Palmers Green allocation and increased frequency as part of restructuring and DMS conversion of RM route 123 which lost its Palmers Green allocation and was concentrated at Walthamstow. Leyton's route N96 was converted from RML to DMS on this date and Upton Park's 169A was withdrawn completely. Over at North Street, route 175 lost its RTs for a second time but, on this occasion, the replacements were standard RMs. The 21st May changes, which included the end of crew work on route 51, also witnessed a big drive towards eliminating RTs from their last stronghold in south-east London with allocation of RMs to routes 161, 228 and a revamped 229 at Sidcup, and to route 161A at Abbey Wood.

Top Sutton received RMs for route 164/A in January 1977, four months later than originally intended, gaining vehicles such as RM 449 ousted from Peckham on receipt of new Metropolitans. Alan B. Cross

Centre Amongst the vehicles received at Sidcup in January 1977 for the changeover from RT to RM of route 51A was RM 486. It is doubtful if the effort was worthwhile as this route disappeared entirely in May 1977 at the same time that route 51 was converted to DMS. J.H. Blake

Left Express operations superimposed upon conventional stopping services have seldom been a success, gaining too little in journey time to be worthwhile whilst proving a source of annoyance to intending passengers at intermediate stops. The longest lived, North Street's 174 Express, was withdrawn on 23rd July 1977. RM 114 is seen with its blue 'Express' blinds on the last day of operation. John Reed

Above **The employment of Fleetlines as crew worked buses on busy routes proved an expensive mistake, and the first route to revert back to RMs was the 29 in March 1977. Holloway's RM 212, despatched from overhaul some three months earlier with registration transfers for the fleet number because of a temporary shortage of the correct ones, sets out from Wood Green garage. Left behind is DM 1192, one of the vehicles transferred in disgrace from the 29 to the 141.** Capital Transport

RMs displaced from the 51 group and elsewhere in May 1977 had permitted the conversion of a whole host of former RT operations at Sidcup on routes 161, 228 and 229. Of these, the 228 was further converted the following January to DMS. RM 678 with a rather battered front dome, is seen at Chislehurst. Capital Transport

1977 will be best remembered in the Routemaster context as the year of the Silver Jubilee of Her Majesty Queen Elizabeth II, and the fleet of silver liveried SRMs which celebrated it are described in another chapter. As the months passed, all of the non-scheduled RT allocations which had bolstered up the Routemasters gradually disappeared as Certificate of Fitness expiries took their toll. September saw the full reversion to RMs of route 230 and the last 'odd' one of all, at Holloway, was withdrawn early in December. In November, the GLC gave approval for the purchase of 250 Leyland Titans and 200 Metrobuses, the latter to follow five ordered in July for evaluation purposes, setting the scene for the nineteen eighties when the two models, together with the Routemaster, would form the backbone of the fleet.

By the start of 1978, a serious shortage of RTs was affecting their ten remaining services and roles were now reversed with RMs deputising for RTs on a frequent basis. To ease the problem, Enfield's route 135 was unexpectedly converted to RM midway through the day on 16th January but this gave only a minor respite. Early delivery of DMs enabled a start to be made on converting Brixton's share of route 109 to doored bus operation on 1st February, providing RMs for early conversion of Palmers Green's route 102. This garage still operated RTs on route 261 and was scheduled to do so until conversion to DMS omo on 22nd April but RMs took over here also and the last RT was delicensed on 1st April. Bromley and Catford shared route 94 which, unofficially, became predominantly RM worked during February and continued to be so throughout March and most of April. At Kingston, route 71 received its first RMs direct from overhaul on 16th February and by 3rd March this route, too, had succumbed. On top of this activity, a series of service revisions on 29th January saw Sidcup's route 228 passing from RM to DMS operation whilst westwards, at Hounslow, a major revamping saw the previously BL operated 237 now with RMs whilst route 117 lost its RMs in favour of Leyland Nationals. A few ex-LCBS RMLs were now arriving to improve Routemaster availability but the main benefit of these was not to be felt until 1979.

The chronic shortage of serviceable buses which had decimated service reliability throughout the mid-seventies continued in 1977. Of the vehicles visible in Catford garage forecourt only RT 2410 is plated up and fit for service. The remainder, including RM 844, RM 982 and SMD 430 in the row nearest the camera, have all broken down. So many defunct buses in full public view hardly constituted a good advertisement for London Transport. Ken Blacker

Centre **Dotted across the fleet were buses of various types which, having broken down, proved impossible to repair promptly through lack of spare parts and subsequently fell prey to cannibalisation in order to keep others going. One of the most extreme Routemaster cases was RM 931 which was delicensed at Peckham in July 1973 and remained off the road right through to January 1981. When photographed at the back of Bexleyheath garage in 1979 it had lost nearly all of its mechanical and major electrical units, almost all cab fittings including the door, some windows, seats, and much else besides.** Joel Kosminsky

Right **The most unconventional conversion to RMs occurred on route 135 when vehicles were scraped together between peaks on 16th January 1978 to enable Enfield's last RTs to be immediately despatched to cover serious shortages elsewhere. RM 320 stands outside the garage which, although always officially referred to as Enfield, is correctly shown as Ponders End on destination blinds.** Frank Riley

'Busplan 78' was the title of a wide ranging review of services aimed at simplifying route patterns, co-ordinating frequencies over common sections of route, and standardising times on wider headway services for easy remembering. Originally intended for implementation in two stages, the first in April and the second in October, it actually required a third in March 1979. The first stage on 22nd April 1978 brought about a reduction of 87 in vehicle requirements, thereby considerably easing the NBA position. On the RM front, Walthamstow's route 256 was completely withdrawn, New Cross routes 151 and 192 and Palmers Green's officially RT but actually RM route 261 were converted to DMS omo, and West Ham's route 262 was handed over to DM vehicles culled from cuts elsewhere. RMs took over yet another of the remaining RT routes, Plumstead's 122 and, for a while, sufficient RTs were now available to enable route 94 to revert fully to its correct rolling stock. On 29th April, Upton Park's route 101 was converted to DM, a move originally planned for February, releasing RMs for Shepherds Bush and Southall to oust RTs from route 105. By 9th July, the DMs received at West Ham for route 262 had found their way on to route N99 in place of RMLs.

It was not long before Bromley and Catford again found themselves short of RTs for route 94 and relying ever more heavily on RMs, such was the high rate of RT ticket expiries, but the situation was again rectified on 15th July when, at short notice and a month ahead of schedule, RMs went to Harrow Weald in place of RTs on route 140. These were the garage's first RMs, received fifteen years later than originally intended after Union opposition had stopped the plan for Harrow Weald and Edgware to start off the post-trolleybus conversion Routemaster programme. Buses from Harrow Weald enabled route 94 to revert again fully to RT, but only until 27th August when RMs officially took over, having been made available by the arrival of DMs on Thornton Heath's 109. This occurred just four days after the last of the much maligned Fleetlines was taken into stock, ending another unhappy phase for London Transport. The only RTs now scheduled for service were at Barking on routes 62 and 87, but the original plan to withdraw them from both in October had to be modified because rebuilding of the railway bridge at Chadwell Heath on route 62 to permit passage of the wider Routemaster had not been carried out. Route 87 became RM operated on 28th October concurrent with the implementation of Busplan stage 2.

Above Left **To speed the departure of RTs from Kingston's route 71 in February 1978, RMs were drafted in straight from overhaul, giving the route a very smart appearance almost overnight. One such vehicle was RM 1991.** Capital Transport

Left **The end of RTs was hastened when Harrow Weald's route 140 was turned over to RMs in July 1978 somewhat earlier than planned. RM 40 passes South Harrow carrying a yellow destination blind denoting a Sunday-only deviation from the main line of route via Cherry Lane cemetery, incorrectly referred to as 'Cherry Tree'.** Capital Transport

This major scheme saved no fewer than 281 buses and resulted in a fairly large surplus of RMs, with a dozen transferred to training duties and about forty being placed in store in the disused Clapham garage. Even more would have been redundant were it not for the appalling unreliability of the DM vehicles on crew services where RMs were now common-place helping to maintain a semblance of reliable cover. Even garages without RMs received a few, such as Muswell Hill and Upton Park and, most notably, Potters Bar where there had been no Routemasters of any sort since 1973. A particularly interesting DM route on which RMs regularly appeared was Edmonton's 283, which had commenced life as a Fleetline operation at stage one of Busplan and had never had an official Routemaster presence. The 28th October changes saw the end of RMLs at Muswell Hill, Tottenham, and Riverside with the reallocation of routes 13 and 243, and the withdrawal of route 74B. Finchley, the pioneer RML garage, got some of the class back along with route 13, and Houns-low obtained a first-time allocation on route 37. A few spare RMLs went to Bow for route 8A but, on the debit side, Riverside's night route N89 was demoted from RML to RM and Stamford Hill's route 97 was lost entirely.

An autumn memorandum from the Executive to the GLC admitted the enormity of the Fleetline failure in seeking approval, granted on 27th November, to purchase a further 450 Titans and Metrobuses to act as replacements. Furthermore, it stated the problem which now had to be faced following the failure of an intensive Multi-Ride (officially the Universal Bus Ticketing System) experiment in Haver-ing, namely that no ticketing system was available under current technology to spread one-man operation as widely as had originally been hoped. Doored buses on crew routes had proved an expensive failure in terms of slower running, poor reliability and lost patronage and, if a nucleus of crew routes had to be retained, the future of the Routemasters needed reassessment. The report acknow-ledged that "so long as it remained possible to maintain and overhaul Routemasters at an economic price, it would not be economical to replace them with doored two-man buses". The possibility now existed of Routemasters remaining well into the 1990s; indeed some optimists even held the far fetched vision that some might last into the twenty-first century.

Top The introduction of 'Busplan', with its first two stages in 1978, brought about various new routes and ended others. Route 283, effectively a localised northern leg of route 279, proved a failure, but during its short life RMs could often be found on it although the official type was DM. Edmonton's RM 773 was photographed in Enfield Highway. J.G.S. Smith

Centre The Busplan scheme marked the end for RMLs at Riverside in October 1978. As a result scenes such as this, of RML 2627 on night service at Liverpool Street, came to an end. Capital Transport

Right The RTs on route 62 obtained a short reprieve from replacement by RMs originally planned for the same time as route 87 in October 1978, ensuring that a few of these veterans would last into 1979. Early in that year, RT 1989 passes RM 408 in Barking. Capital Transport

The operating management's current concern that as many secondhand Routemasters as possible should be purchased as a safeguard against future requirements bore out the view that many of the class, and certainly all the RMLs, would be around for a very long time to come. Despite an announcement on 13th November by Leyland that it was to close during 1979 the famous Southall factory where many Routemaster units had been built, the engineering department was confident that there would be no insurmountable problems in keeping the RMs going, so sound had their basic structure remained despite the arduous nature of the work they performed.

As a small step towards improving maintainability, a limited programme of modifications was instituted late in 1978 whereby Clayton braking systems were removed from a number of buses and replaced by Lockheeds. The largest programme, embracing 33 buses, commenced with the outshopping from overhaul of RM 1009 on 3rd December and resulted in bodies from the series B1071-1154 classified 9/5RM5/9 being reclassified 12/5RM5/9 to denote the revised brake gear. The spread of bonnet numbers involved ranged between RM 995 and 1581 but most were in the 12 and 13 hundreds. A smaller new sub-class was 13/5RM5/10 which was created out of the 10/5RM5/10 type, of which there had only been sixteen in the first place (B1055-1070). Four of these were dealt with; their stock numbers after overhaul were RM 994, 1129, 1238, 1240. Finally, one RML was converted from Clayton to Lockheed towards the end of the programme in April 1979. This was ex-London Country RML 2315 which still retained its original body but changed from being 2/7RM7/3 to become the one and only 5/7RM7/3. Somewhat earlier than this, back in March and April 1978, two high numbered bodies received non-opening front windows when rebuilt after accidents; these were RM 494 (B1196) at Poplar and RM 1218 (B1166) at Battersea. A modification introduced progressively from 1979 was the fitting of a glass fronted fire extinguisher contained on the rear platform with the removal of the extinguisher from the driver's door.

The early part of 1979, marked by the industrial 'Winter of Discontent' and bitter weather found numerous Routemasters with seized engines and a host of other ailments amongst the 600 or so daily NBAs. Busplan had signally failed to provide the reliability which it had promised, due to a worsening driver shortage which now stood at 13% but, nevertheless,

stage 3 went ahead on 31st March saving a further 74 buses. Most significant was the reconversion to crew operation of route 106, whose reliability had disintegrated under omo, for which Hackney and Tottenham both received additional RMs. The only other major RM related change resulted from discussions with Surrey County Council and saw the conversion of Sutton's route 164 to DMS along with the withdrawal of 164A.

Few tears were shed when the first DMS departed for scrap on 26th February in contrast to the mass nostalgia which came to the fore on Saturday 7th April when, at long last, RMs unseated the last few remaining RTs from route 62. The offending railway bridge at Chadwell Heath which had caused the retention of the RTs had still not been rebuilt to accommodate wider vehicles, but it was impossible to keep the RTs any longer in the face of impending CoF expiries so the route was diverted from the 7th. The morning started off with a number of RTs in service but, in a well managed operation, each was substituted for an RM at Barking on crew meal reliefs except for the last one of all, RT 624, whose final trip was duplicated by RM 208 in order to carry the normal passengers unable to board the RT laden with enthusiasts.

Barking's Gascoigne Estate is the setting on 7th April 1979 as fully laden RT 624 swings past RM 350 during the daytime changeover from one type to the other on the final day of RT operation. RT 624 was destined to make history as the last of its class in scheduled service with London Transport. David Stuttard

On 22nd April 1979, route 24 reverted to RML operation, these having been made available through a direct swap with Stonebridge garage whose crews on route 18 had requested doored buses in the hope of overcoming a particularly bad spate of hooliganism. The summer of 1979 was the time of the Shillibeer and Shoplinker RMs and the failure of the Shoplinker experiment provided RMs to displace DMs from West Ham's route 262 on 14th October and Upton Park's 101 a fortnight later. No new route was found for the surplus DMs which were all absorbed covering for defective buses elsewhere but, in any case, no willpower remained to convert permanently any more crew routes to doored buses in view of the acknowledged failure of such conversions. London Transport, under heavy pressure from the GLC to improve performance after failing to meet promised targets, decided on a large measure of decentralisation resulting in the formation of eight semi-autonomous Districts on 1st October. The individual symbols of these Districts began to appear on buses from 11th December, in the case of Routemasters on the blank window-level panel immediately adjacent to the entrance platform.

After only nine months with RMs, on 13th January 1980 Barking's route 62 became the first on which Routemasters gave way to the new T class Leyland Titans which were fast establishing a foothold in east London. Route 87 followed on 17th February, both being crew worked at this stage but due for early one-man conversion. The RMs were required for Peckham's route 36 group in place of MDs which, it had now been decided, would not be retained beyond their CoF expiry at seven years old because of their non-standard nature. The 36 group had been selected as the pilot for BUSCO, an experimental two-way radio communication system using micro-chip technology and VHF radio to give fast, direct speech and digital communication between drivers and a central route controller. Expensive equipment had to be fitted on over sixty buses and it was thought prudent, in view of the limited life expectancy of the Metropolitans, to replace them with Routemasters. From 3rd February, MDs began to drift into Plumstead garage in what was effectively a swap for the RM rolling stock from route 122 which had become totally MD by the 24th of the month. 10th May saw the closure of Turnham Green garage and its replacement by a refurbished Stamford Brook which, in its former guise as Chiswick Tram depot, had already served as a Routemaster operating base for the BEA coach fleet. The buses showed no sign of the change as the old V garage code was retained.

Three major replacements of DMs by Routemasters occurred during the summer, many of the displaced Fleetlines being converted for driver only operation to replace others destined for sale or scrap. The influx of ex-London Country RMLs was the catalyst which made possible the conversion of Cricklewood's 16/A on 25th May 1980. On Hanwell's share of route 207 began to receive a mixture of RMs and RMLs, a process which was not complete until the end of October though it was to take until September 1982 before the RMs too were displaced to give a 100% RML allocation. Uxbridge's share of the route fol-

Non-standard external features were few and far between on Routemasters until the nineteen-eighties Showbus phenomenon. Streatham's RM 679 was rebuilt in 1979 after an accident with the rearmost upper deck opening window in the wrong bay. *Gerald Mead*

The late 1970s found two standard RMs fitted with plain upper front windows when quarter drops were unavailable to complete repair work, but their appearance differed from the early batch built in this form due to the lack of ventilator scoops. One of the two was RM 1218 seen operating from Battersea with body B1166 which was soon afterwards transferred to RM 1220; the other converted body, B1196, was on RM 494 and later RM 1547. *Colin Stannard*

lowed from 20th August, using RMs released from the training fleet whose holding of standard Routemasters was being drastically reduced. The 207 conversion was, however, overshadowed by events in north London where, also from 10th August, the first RCLs began to enter bus service in red livery. Their intended route was the 149 and Stamford Hill was the first recipient with Edmonton's RCL operation following from 8th October. Although officially allocated to route 149, both garages actually placed the RCLs freely on all their Routemaster operations making them commonplace on routes 243, 253, 279 and 279A. A large number of changes within Leaside District on 27th September saw the end of RMs on routes 26, 135 and 298/A. Finchley's

26, much extended and converted to omo, was the first Routemaster service to be handed over to Metrobuses. Enfield's route 135 became DMS, as did Palmers Green's 298 which was also extended and reallocated to Potters Bar and Wood Green. Route 298A was lost entirely among other changes. On the same day, New Cross route 192 ceased as an RM operation in favour of MDs from Plumstead. The final DM to RM reallocation of the year involved route 168 on 15th December and the night services of both participating garages (N68, N81, N87 and N88) also officially became RM although only at Wandsworth on routes N68 and N88 did this actually happen, Stockwell preferring to use DMSs or Round London Sightseeing Tour DMs.

At the beginning of 1980 the training bus fleet, shorn of most of its RTs, included no fewer than 73 standard RMs as well as examples of the four other Routemaster classes totalling a further 77. Standing by the skid patch at Chiswick, with its red flag denoting that skidding is in progress, is RM 1187 bearing the garage code for Bexleyheath, one of the few garages never to operate Routemasters in public service. FRM 1 is also present on one of its periodical visits back to Chiswick. Conversion of Fleetlines to trainers later permitted most standard RMs to revert to their normal duties. Ken Blacker

1980 saw the birth of the Showbus phenomenon whereby staff adopted a bus to adorn, in their own time and expense, for display at rallies and other events. Probably the first were Sidcup's RM 704, Catford's RM 1250 and Croydon's RM 1000 but others quickly followed. Encouraged, or at least not resisted by management because of the morale-building value, the Showbus craze resulted in many restorations to various stages of early condition, including the refitting of full depth front heating grilles, original style radiator grilles and brake cooling apertures and, in some cases, even the reinstatement of the offside route number indicator. Some interesting points of attention to detail emerged, including the replacement, in at least one case, of the newer style yellow lower saloon bell cord with an original burgundy one. Inevitably some, like the polished brass lamp surrounds of RM 1644, went completely overboard. One common point was the almost universal reinstatement of gold fleetname transfers. This coincided with the very last 'genuine' survivors still to carry them which were, not surprisingly, now looking very shabby. The final RM to carry the traditional underlined fleetname, temporary trainer RM 641, had disappeared into overhaul in August 1987 but RML 2539 and RM 1128 were still active with the block capital fleetname into 1980. RML 2539

The Showbus phenomenon, which began in 1980, produced some splendid labours of love. It unleashed enthusiasm in garages which had been latent, even unsuspected, during the old rigid regime but which was now finding greater freedom under the new district control. Mortlake's RM fleet was always immaculately turned out, but pride of place went to Showbus RM 1563 which positively sparkled. With polished wheel and lamp trims, a restored offside route number, old style gold lettering, but a confusion of both old style bonnet motif and later type grille, it is seen at work on route 9. RM 737 is seen stripped down by Harrow Weald staff to reveal the old framework. Capital Transport

A sigh of relief was probably given by staff and passengers alike when route 16 reverted to RML in May 1980 after its unhappy flirtation with DMs. Prominent amongst the new intake at Cricklewood were many newly overhauled ex-London Country vehicles such as RML 2445. *Capital Transport*

survived at New Cross in this form until June 1980 whilst RM 1128 at Peckham lasted through to April 1983. While the local enthusiasts were busy restoring Routemasters to various stages of original (and sometimes unoriginal) glory, London Transport toyed with the idea of reducing the standard of appearance by eliminating the white band on economy grounds. This was not an original idea, having been tried on a few buses at Highgate many years earlier, but its revival was logical in view of the latest policy of spraying all other types in unrelieved red. RM 376, outshopped from Aldenham in early December, was examined at Chiswick but fortunately the powers in charge baulked at debasing the standard for Routemasters and the white band was restored prior to entry into service.

1981 was quieter than usual in terms of revisions to London's bus service network but in other ways it was one of the most notable, even traumatic. No fewer than three new garages were opened after a lapse of more than a quarter of a century, and a defunct one re-opened; on the debit side, six closed for good. The last Merlins and Swifts, whose very purchase in such large numbers had once posed a threat to the future of the RM family, were withdrawn in July to close a very unhappy chapter which had cost London Transport dear. To cap it all, a new GLC administration, elected in May on a manifesto of reducing public transport fares by 25% in fact did so under its famous 'Fares Fair' scheme of 4th October, stimulating a massive increase in

ridership, a welcome drop in private motoring, and an unwelcome legal challenge from the London Borough of Bromley culminating in a House of Lords judgement confirming the fares drop as illegal and leaving London Transport in a state of total confusion at the year's end. On a lighter note, the wedding of Prince Charles and Lady Diana in July was celebrated with a small fleet of specially decorated RMs.

The now well established policy of removing doored buses from crew work on busy London services was taken a stage further between 23rd January and 26th February as the DMs and MDs working from New Cross on route 53 were replaced by RMs. Further progress in the same vein on 31st January saw RMs taking

over Holloway's route 4 from DMs. Riverside's route 72 lost its RMs on the same date when one-manning brought in Fleetlines worked by Shepherds Bush, Riverside garage being physically unsuitable for the larger vehicles. Tottenham garage gained a few RMLs for a newly acquired share of route 243 whilst a complex scheme in Forest District resulted in Leyton-

Royal wedding day on 29th July 1981 was celebrated by London Transport with the small fleet of specially painted Routemasters described in chapter 10. A local effort, probably by the crew of this Croydon-based bus, was the Union Jack sported by RM 1171 at Holborn Circus on one of the many route diversions necessary that day. *Capital Transport*

London Transport's most primitive garage by far was Middle Row which, unable to accommodate Fleetlines and other new type vehicles and unsuitable for rebuilding, closed down upon the opening of the new Westbourne Park on 15th August 1981. RM 239 stands at the rear of a queue of buses waiting to run in shortly before the closure. Capital Transport

One of many services converted to driver-only operation in the September 1982 cut-backs was Palmers Green's 102 on which RM 1570 is seen at the Golders Green terminus. This bus was amongst the first large batch of Routemasters to be sent for scrapping, passing to W. North, the Leeds breaker, in October 1982. Alan Nightingale

worked RM route 230 being converted from RM to LS omo whilst its route 55 was dealt with in the opposite fashion by going from DMS to crew RM, both routes being subject to modification and extension at the same time. Also at Leyton, route 48 reverted from RML to RM to permit the longer buses to cover a proportion of the total requirement for route 38. Starting on 5th February, Abbey Wood's share of route 180 was progressively converted to MD, still crew worked, whilst on 21st March, Walworth's allocation on route 45 became RM once again.

25th April 1981 was noteworthy as the opening date of the palatial new Ash Grove garage, marking the closure of Dalston and Hackney. Amongst the vehicle intake were RMs and RMLs but the takeover was not straightforward and there were various route reallocations which included the end of RMLs at Stamford Hill with the loss to Tottenham of route 243. Still on 25th April, the process of eliminating Holloway's large DM holding, which had commenced with route 29 and later route 4, was progressed to completion with the placing of RMs on routes 45 and 172 (with Routemasters, RM or RML, taking over routes N92 and N93). Over at West Ham, route 5 was converted back to crew operation using RMs

(Poplar's share of the route being discontinued) but, at the same time, route 262 received DMSs in place of RMs on conversion to omo.

Another auspicious event on 25th April was the reopening of Clapham garage which became operational for the first time since 1958, having served in the intervening period as the much loved Museum of British Transport and later as a store for unlicensed buses. Planning consent for its conversion back to its original use was reluctantly granted by the local authority for six years to enable the garages at Norwood and Streatham to be demolished and new ones constructed on their sites. The old Norwood closed first and its crews, though not all of its routes, transferred to Clapham, retaining their N code letter on the bus sides even though the garage was officially coded CA. Unlike Norwood, Clapham had an allocation of RMLs as well as RMs, having acquired Stockwell's share of route 37, and it soon began putting its distinctive trademark on both types by painting the radiator grilles silver. As part of a fairly wide ranging revision within Wandle District on 25th April, Stockwell and Wandsworth garages lost RM route 168 which ceased to run.

The opening of the new Westbourne Park garage on 15th August 1981 meant the absorption of all of Middle Row's totally RM

work plus one RM route from Stonebridge, the 18A, along with most other workings from that garage when it closed. Westbourne Park was controlled by Abbey District and, to avoid bringing Metrobuses into the district where none already existed, a complex shuffle was arranged whereby Stonebridge's Ms replaced Fulwell's RMs on routes 33 and 281 which in turn travelled to Brixton to depose DMs from routes 109 and 133 in order to make them available for Westbourne Park. Upon transfer to Westbourne Park, route 18A was officially converted to DM but, in practice, their appearance was spasmodic and RMs remained. By this time, the fairly rigid adherence of nominated bus types to routes was, in any case, tending to break down under the new district structure with frequent appearances at mixed-type garages of DMs, Ms and RMLs on RM routes and vice versa. One new route was inaugurated at the Westbourne Park opening, the 52A which was RM worked.

Finally in 1981 came the opening, on 31st October, of Plumstead garage with the consequential closure of Abbey Wood and the old Plumstead premises. The new garage started life with a hundred percent allocation of Metropolitans on both crew and driver-only work. Abbey Wood had retained a few RMs to the end for route 161 whose workings were taken over by MDs although Sidcup's share remained RM. The old Plumstead had no Routemasters at the end. Under the new regime, the whole of the 180 was a Plumstead responsibility and its minority Catford RM operation was withdrawn.

Very visible non-standard modifications were made to a number of Routemasters during overhaul in 1981 because of temporary non-availability of the preferred materials. During March and April, some 54 RMs appeared with burgundy coloured seat backs (the same as was fitted to the RCLs) in place of the usual Chinese green, which actually suited their appearance. Vandalism to rexine covered seat backs was a particular problem and, soon afterwards, plastic seat backs were devised for fitting on an 'as and when' basis although these never became as widespread as originally intended. During June to August 1981, half a dozen RMLs (RMLs 2609, 2612, 2617, 2624, 2637 and 2649) were retrimmed throughout in the popular but aesthetically unsuitable RT moquette followed in September by Sidcup 'Showbus' RM 770 which was the reincarnation after overhaul of their earlier display bus RM 704. More significant in 1981 were the invisible alterations which started in February when RM 227, carrying early style body B217, was equipped on overhaul with new and improved types of CAV manufactured alternators, rectifiers and control panels, and was given the classification 14/5RM5/11. This was followed from July onwards by a steady programme of re-equipment which predominantly touched on bodies numbered below B500 but with a number of exceptions, one of which was float body 9986 on RM 397. Both CAV and Simms equipped buses were dealt with, being linked together for the first time under the 15/5RM5/11 banner or, in the minority of cases which had SCG rather than Lockheed transmission systems, as 15/5RM5/12. A further code, 15/5RM9/3, is believed to have been unique to RM 2115 which, in carrying body B2035, was the only illuminated advertisement bus to be dealt with. About 47 RMs were converted during 1981, a roughly similar number throughout 1982 and a few more in 1983 to give a total of about one hundred. Some 40 RMLs were also re-equipped with the same CAV material from March 1982 onwards but, in their case, only those originally with Simms equipment were slotted into the programme. New coding categories applied to these conversions were 6/7RM7/8 for buses with Lockheed brakes and 7/7RM7/10 for Clayton-Dewandre equipped buses. No differentiation for transmission systems was necessary as all RMLs were SCG equipped; likewise the RMs needed no separate categories to denote braking systems as all those converted were Lockheeds. Additional categories set aside for the RML conversions were 6/7RM7/9 and 7/7RM7/11 which would have been used for illuminated advertisement bodies but, in the event, these were not required. One of those converted to 6/7RM7/8, RML 2376 carried body B2315 which, in 1979, had been converted into the unique 5/7RM7/3, a category which was now discontinued.

Following the Law Lords' unexpected and much criticised December ruling that Fares Fair had been illegal, 1982 emerged as a tragic year in the history of London Transport.

Fares had to be doubled from 21st March, bringing the unavoidable spin-offs of a massive decrease in ridership, a reduction of 15% in scheduled bus mileage, restrictions on recruitment and overtime levels, and subsequent job losses, industrial disputes by both operating and engineering staff, the prospect of inevitable garage closures and, finally, the first mass withdrawal for sale or scrap of Routemasters as the proportion of crew work in the fleet was speedily reduced to 52% of the total in order to save money. The buoyancy of 1981, which proved that low fares and attractive service levels were the key to keeping a major city on the move, was transformed almost overnight. The GLC mounted a vigorous Fare Fight campaign but the Government failed to heed it, preferring to discuss removing London Transport from GLC control. Nor did it heed a House of Commons all-party Select Committee recommendation to set up a Metropolitan Transport Authority with rate levying powers formed through a liaison between the Department of Transport, the GLC, and the London boroughs. By the end of the year, 98 standard RMs were in the process of being scrapped at Aldenham and a further sixty-odd had been sold as part of the withdrawal of some 580 buses from the fleet as a whole. In selecting RMs for disposal, those with Leyland engines or Simms electrical gear were regarded as the first priority.

Little happened to affect the Routemasters in the first half of the year, except for the commencement on 25th March of their replacement from North Street's routes 174 and 175 by new crew-operated Titans and subsequent restoration to route 141 in place of DMs. The changes were gradual; North Street's last Routemasters did not depart until October whilst Wood Green retained DMs until June, and New Cross until September. On 4th June, the conversion of route 172 from DM to RM, commenced in 1981 with Holloway's allocation, was completed when Camberwell's share followed suit. The massive service cuts, which everyone had been bracing themselves for, were scheduled for 31st July but subsequently deferred to 4th September to accommodate protracted union negotiations. Many reallocations of services from one garage to another took place but the changes most notably affecting Routemasters included the total withdrawal of routes 18A, 94 and 176A which, in numerical terms, was more than balanced by new routes 60 (worked by Croydon), 208 (Catford), 225 (Seven Kings) and 261 (Bromley), all worked by RMs. RMLs made their first appearance on Ash Grove's share of route 11 and their partial allocations on routes 38 and 207 (Hanwell's share) were completed. Services restored from DM to RM were 43 (Muswell Hill), 109 (Thornton Heath; Brixton had been dealt with earlier) and 134 (Potters

April 1983 saw removal for the time being of double deck operation from Bromley's route 261 when Routemasters were usurped by new Leyland Nationals. RM 1547 heads down Mason's Hill into Bromley town centre in the previous September. J.H. Blake

Bar and Muswell Hill) whilst Peckham's route 63 changed from MD to RM. On the debit side, crew-worked Ms displaced RMs from Alperton's route 83 and Southall's route 105 (although a few RMs remained until December) and DMs went to Sutton for route 93. One-man conversions removed Routemasters from routes 101 (Upton Park), 102 (Palmers Green), 104 (Holloway), 106 (Ash Grove) for the second time, 187 (Alperton and Westbourne Park), 193 (Seven Kings), 196 (Stockwell), 229 (Sidcup), N89 (Riverside), and N99 (West Ham). Finally route 180, which for a while had been an all-Metropolitan allocation from Plumstead, regained RMs at Catford with the resumption of a localised operation. Garages which lost their RMs completely as a result of these changes were Alperton, Southall and Sutton, plus Norbiton which lost route 65 to Kingston. With more than one thousand bus movements officially to be made overnight, some of the reallocations were staggered with, for instance, Muswell Hill receiving RMs ahead of the planned date. The massive scheme reduced scheduled Monday to Friday Routemaster requirements by 241 (219 RM, 17 RML, 5 RCL) and left just three garages with one hundred per cent RM allocations: Kingston, Mortlake, Riverside.

By the year's end, the programme for fitting the whole fleet with two-way radios was almost complete and, in addition, a few Routemasters were now equipped with tachographs to enable them to undertake private hire work within the current regulations. On 22nd November, West Ham's route 58 was converted from RM to crew-operated Titan to compound the Routemaster's blackest year to date.

The see-sawing of fortune continued into 1983 when High Court approval was given for a 25% reduction in fares in conjunction with an integrated Travelcard scheme which all came into effect on 22nd May. This was London Transport's Golden Jubilee year and its special efforts to celebrate it had less heart than would have been the case under more stable circumstances. Legislation to transfer the control of London Transport from democratically elected representatives of London to a Government department met with fierce opposition, particularly from all parties on the GLC, but it pursued its course through Parliament. A massive programme of omo was instituted on 23rd April, encompassing thirteen routes although RMs were directly affected in only four of them. The remainder had already been taken over by doored buses as part of a recent policy of pre-positioning the new types and letting them run in before conversion. The four routes retaining RMs to the end were Croydon's route 60 which gave way to DMSs, Seven Kings' route 225 taken over by Ts, Bromley's route 261 taken over by LSs and, finally, Harrow Weald's route 140, some of whose new Metrobuses had already put in an appearance alongside the declining RMs. On the same date, Finchley received a small allocation of RMs (it already had RMLs) to cover new workings on route 43, but Potters Bar lost its RMs when route 134 was shortened to terminate within the GLC boundary at Barnet, the section to Potters Bar being covered by an extension of omo route 263. These were the last omo schemes of 1983, the trade union

Seven forlorn looking Routemasters await their fate. They are among the 150 or so stored in Ensign's Purfleet yard after the September 1982 withdrawals. They are, from left to right, RMs 1706, 1379, 1578, 82, 865, 420 and 358 of which all except the two nearest the camera are Leyland engined. Some were scrapped at Aldenham, others at North's. J.H. Blake

The one-time Thames Valley garage at Uxbridge was inconveniently located and by the time of its closure in the early hours of 3rd December 1983 had become badly inadequate by modern standards. RM 337 was captured on film there a couple of years before closure. J.H. Blake

refusing to participate in any more, so the RMs were given a respite. Not so, however, the RCLs whose gradual withdrawal commenced in July although most were still operational at the year end.

On 25th June 1983, two of London Transport's smaller garages, Mortlake and Riverside, closed. In the case of Mortlake, closure came only after a bitter struggle and its demise meant the loss of the last garage to take real pride in the appearance of its buses. These displayed, to the end, the standards which had been the expected norm throughout the fleet in London Transport's better days. To an extent, Mortlake had been helped in this respect by the fact that, through being fitted with a special radiator filling system with

extra pipework etc, its RMs tended to stay put more than at most garages; some Victoria RMs were similarly equipped. Mortlake's work was divided between Stamford Brook and Fulwell, with the result that route 33 which it had shared with Fulwell's Metrobuses now lost its RMs, whilst Riverside's operations passed to Victoria and Shepherds Bush. Under the harsher financial criteria now imposed, the individual districts were experimenting with initiatives one of which, under the aegis of Cardinal District, saw the introduction on 12th November of two new Kingston-area shoppers' express services, the K1 and K2. They worked on various days of the week from Kingston garage using RMs with special yellow blind displays.

The Metrobuses and Titans flooding into the fleet from 1979 onwards were equipped with side blinds combining route number and destination which drivers, at the behest of their union, refused to set correctly for each journey. In a moment of capitulation they were withdrawn and many were subsequently re-used where appropriate as destination screens. Alan B. Cross

The GLC, never slow to involve London Transport in its political machinations, made RM 1152 the focal point of a poster campaign condemning the Transport Bill. This was one of the battles against the government which it was destined to lose. Colin Stannard

September 1983 saw the start of an advertising campaign by the Midland Bank employing large L-shaped advertisements on the sides of Routemasters which, though ugly and unsympathetic to the lines of the vehicle, were an omen of things to come. RM 1717 had earlier been transferred to London Transport Advertising for demonstration purposes in connection with this scheme and remained with them for further trials. 3rd October saw the live introduction of the BUSCO trial on routes 36/A/B for which 71 of Peckham's RMs, including a pair of prototypes (RM 696 and 2051) had been modified at Camberwell garage. The on-bus computer equipment was housed in a protective steel case below the upstairs front offside seat which was raised by about one inch to accommodate it. The running number reader units were sandwiched between the inner and outer offside panels behind the normal running number plate holder which was lowered by a few inches and provided the easiest means of identifying a BUSCO bus. For some three years prior to the BUSCO introduction, RM 45 had been used for experimental purposes in connection with it and had acquired platform doors. It never returned to public service.

The 1984 of George Orwell's predictions turned out to be the year of reckoning for London Transport. The year started off in an air of contraction as the number of operating districts was reduced from eight to six on 2nd January. Twelve days later, Kingston garage

closed and its Routemasters on routes 65 and 71 passed to Norbiton. However, like Riverside, Kingston stayed open for operational purposes, in this case as a bus station. An already depressed morale was not aided by an ill-timed management decree that all Show-buses, of which there were now a good number, must immediately revert to standard livery. Not so long earlier, during the 1983 season, Cricklewood's Showbus RM 383 had been given the RT style seat cushions and squabs formerly on RML 2649 whilst RM 8 at Sidcup had similarly received the seats from RM 770.

The final positive act under the old regime was the introduction, on the night of 13th/14th April, of a much expanded night bus network which quickly proved itself immensely popular. However, with the exception of routes N29 and N93, it marked the end of crew operation on night services and though all had finished their days, officially or unofficially, with doored buses for safety or capacity reasons, Stamford Hill marked the last crew night by operating RCL 2251 on route N83 whilst Stockwell put RML 2389 out on route N87. Most of the remaining effort in the first half of 1984 consisted of replacing RMs with newer buses on routes intended for one-man operation at a not too distant date. Seven Kings' route 86 started off by receiving Ts from 12th February although the RMs proved tenacious and made appearances up to July. Titans also replaced Sidcup's RMs from route 161 from 24th April and Poplar's on route 40 from about the same time, although here again RMs were evident in declining numbers until July. Willesden's route 260 received Metrobuses from 17th May, whilst Titans took over Walworth's route 176 overnight on 11th June. Still more Titans enabled a progressive takeover of Catford's route 208 from 11th June and of West Ham's 69 a few days later.

The London Regional Transport Act received Royal Assent on 26th June and it was known to be only a matter of time before financial and policy control of London Transport passed to the Secretary of State for Transport, Nicholas Ridley. However, the sudden unannounced nationalisation on 29th June took everyone by surprise, not least the GLC who, if they had plans to thwart the takeover, could not now implement them. Less than a year after celebrating its Golden Jubilee, London Transport as we knew it ceased to exist. Under the new LRT (London Regional Transport) which took its place, life was to be very different. With a government-directed commercialism involving massive subsidy reductions and competitive tendering for bus services, the opposite approach to public transport provision from that of the GLC would prevail. Matters had been less than satisfactory under the GLC with its changes of political control and direction every four years preventing any semblance of long term planning. The final administration's eagerness for confrontation with Whitehall and provocative measures like packing the London Transport Executive with appointees of little obvious qualification, had sadly detracted from its very real concerns for public transport. Under the strict financial regime of LRT, a much enhanced omo programme and a drastic speeding-up of Routemaster disposals was a very real possibility.

A major change of ownership has taken place, but outwardly little has so far changed on Romford's RCL 2224 except for a new name and address on the legal ownership panel. In true Green Line tradition this is in neat gold transfers. When this photograph was taken early in February 1970 the Company was in the process of removing the London Transport bullseyes from their prominent mid-decks position.
Alan B. Cross

CHAPTER TWO

LONDON COUNTRY

In November 1969, five weeks before ceding to the National Bus Company, the staff of Country Buses and Coaches were given the first sketchy details about the "New Era" which lay ahead for them. They learned that they would be "part of a very large transport company . . . which has 67 subsidiary companies throughout England and Wales, with 27,000 vehicles and over 80,000 staff". They discovered that they were to have a new Managing Director, Mr C. R. Buckley (whose very long career had been mainly with Crosville), and a new fleet title London Country. A buoyant picture was presented to a mainly sceptical staff but in truth all was far from well and a long uphill battle lay ahead during which the Routemasters were destined to be withdrawn, looking for the most part distinctly run down if not battle scarred.

On 1st January 1970 London Country Bus Services Ltd inherited, free of charge, assets with a book value of £3,369,000 in its role as a wholly owned subsidiary of the National Bus Company and set up its head office in Reigate, once the seat of power of the East Surrey Traction Company. It possessed an unnatural shaped operating territory likened to a Polo mint (circular with a hole in the middle) with an ailing Green Line network linking parts of the territory via the 'hole' which was central London. It possessed no overhaul or major repair facilities, nor even its own driver training unit which had to be established quickly. The fleet, which was entirely inherited from London Transport, presented a major problem

in itself. Of 1,267 vehicles no fewer than 775, or 61%, dated from the 1950s or earlier. Even more significant than the inadequacy of capital investment by the old management was the inappropriateness of much that had taken place, manifested by the fact that, of the 721-strong double deck fleet, only eleven vehicles (a mere 1.5%) were capable of being one man operated. So dire was the situation that, even before the new undertaking became operative, provisional plans had been drawn up for 90 new Green Line coaches and 90 new rear engined double deckers to be obtained as early as possible, which effectively meant 1971.

The two largest classes of vehicle inherited by London Country, both obsolete, consisted of 484 RTs and 263 RFs but following closely behind the RFs came the three types of Routemaster which totalled 209 in all. These were by no means evenly dispersed throughout the fleet, being present at only 17 out of the company's 28 garages. The eleven garages without Routemasters were Amersham, Chelsham, Crawley, Dartford, Dorking, Leatherhead, Luton, St Albans, Staines, Swanley and Tring. The initial allocation of Routemasters to garages, as inherited from London Transport on 1st January 1970 (including vehicles temporarily delicensed or in works) was as shown in appendix 3.

In addition, RMC 1464 was garaged in the London Transport premises at Riverside as part of the longstanding Green Line practice of basing one or two vehicles in central London to cover for breakdowns and other emergen-

cies. For this reason, special blind displays were fitted to permit operation on routes throughout the network. Under the old regime, routine maintenance on RMC 1464 had been carried out at Riverside but this arrangement lasted for only three weeks into the London Country era; on 23rd January the vehicle was allocated to Guildford for maintenance purposes with Riverside serving henceforth only as a parking place.

Not much time was lost in removing the London Transport fleet names and replacing them with the new title in gold capital letters using a plain style without underlining. Being in the famous Johnston style of typeface, the new name gave the impression of having been designed by London Transport, which was probably the case as the first three vehicles to display it – the Blue Arrow XFs at Stevenage – had actually done so in London Transport days. Later in 1970, the Company's own individuality began to show through when the relief bands on some RMLs were repainted yellow and a new yellow motif made its debut on the offside staircase panel. This clever and distinctive emblem consisted of a circle with several horizontal bars representing London Country's ring of green countryside around the capital and also symbolising a wheel and movement. The treelike arrangement of the horizontal bars also suggested the rural nature of the network. It was a great shame that this clever design was to be abandoned within two years by the NBC's imposition of a corporate image policy.

The original LONDON COUNTRY gold fleet names were not very prominent and often blended into dull paintwork almost to the point of obscurity. Radiator badges were turned round to present a plain green image as on RML 2440. Alan B. Cross

In spite of its new name and ownership, London Country continued to display its London Transport origin by retaining the old fleet number series and other outward manifestations such as painted garage codes and running number stencil plates. With no overhaul works of its own, the new organisation was still dependent upon Aldenham and close co-operation continued to exist on the engineering side. If, for instance, modifications were deemed necessary to London Transport's fleet of Merlins (as inevitably they were!) details would be distributed to London Country garages as well as LT's own. Thus when, in November 1971, London Transport decided to eliminate the distinctive rear wheel discs from classes fitted with them including the Routemasters, the same instruction went to London Country garages who mostly complied quite quickly. This particular move, though officially on the grounds of reduced maintenance, was rumoured to have come about after a disc had come adrift causing injury.

Some foreseeable problems lay ahead, a major one being that the Certificates of Fitness of all 43 RCLs were due to expire within a two-month period from May to July 1972 followed by all 97 RMLs between September 1972 and May 1973, presenting a severe challenge to an undertaking with no major repair centres of its own. Only the RMCs provided any respite; RMC 4 had emerged from its second overhaul in May 1970 and the remainder of the class were certified until between May 1974 and February 1975. To

avoid the chaos of a last minute bottleneck, the RCLs began to enter Aldenham for their first overhaul starting on 1st January 1971. By and large, each RCL owning garage was dealt with one at a time starting with Romford, and then in the following sequence: Grays, Dunton Green, Windsor and Godstone. Five of Hatfield's RMCs (1469, 1482, 1487, 1495, 1498) were used as cover for them, joined later by RMC 1455 from Addlestone, and these went around each garage in turn, keeping more or less together, for the twelve-month duration of the programme. Their use in this capacity virtually marked the end albeit temporarily, of RMCs on full time bus work at Hatfield, their role on route 303/A being gradually whittled away during the first three weeks of January as one after the other moved eastwards to Romford. RTs returned to the 303/A but only temporarily; on 2nd February 1971 the service was converted to one man operation with new SM class AEC Swifts ordered in London Transport days but destined to become a prominent, if unhappy, feature of the London Country scene. The temporary RMC allocations at Romford and Dunton Green recalled earlier short term operation of this type but, for Godstone, the RMCs which arrived from 2nd December 1971 onwards were its first. As for the RCLs, although overhauled at Aldenham they were not subjected to the traditional London Transport system of changing identities; each vehicle spent about six weeks in the works and emerged with its original body and running units.

For the first two years of London Country ownership, very little happened to impact upon the Routemasters, a far more pressing need being to whittle away the large but obso-

lete RT and RF fleets and the equally aged but much smaller specialised RLH and GS classes. The only new Routemaster allocations were from 20th February 1971 on route 341/B (with Hertford worked journeys on 310, 395/A and Hatfield trips on 340) although a shortage of RMCs meant that RTs continued to predominate temporarily, whilst on 25th September 1971 High Wycombe's route 326 gained RMLs rendered surplus by reduced requirements at Godstone. However, this quiet period was the lull before the storm which was to strike at the Routemasters with intense fury from the very first day of 1972, changing entirely the role which the whole RMC class and most of the RCLs were to perform within the fleet. The new management had quickly decided that double deckers were wrong in both capacity and image for Green Line operation, and the new, 90-strong, RP class of Park Royal bodied AEC Reliances was ordered to transform the network. On 1st January 1972, the first batch of these was placed in service on route 721, completely ousting Romford's RCL fleet which, at fourteen vehicles, was itself a mere shadow of the intense double deck operation which this garage had once provided. At the same time, Grays received the earlier RC class of Reliances displacing RCLs from route 723 and also from a short-lived 723A, dating only from 20th February 1971, which now ceased as a separate entity. A total of eight RMCs and twenty RCLs had reached the end of their Green Line career and were officially reclassified from coaches to buses. Those based at Grays remained there, the RMCs to continue modernisation of route 370/A commenced in London Transport days, with the RCLs scheduled for routes 300 and 328A/B, although interworking at Grays which was traditio-

nally more intense than at other garages as a result of the takeover from Eastern National back in 1951, resulted in numerous Routemaster journeys on routes 367, 368, 369, 371/A/B and 374. Romford's RCLs departed to Reigate and Dorking for route 414, with Reigate worked journeys on routes 406 and 424, those on the 406 being restricted to specified running numbers which did not operate north of Epsom. Vehicles reclassified from coach to bus quickly became apparent in their new capacity as fleet names were changed, commercial advertising was applied and, in some cases, the pale green band was repainted yellow.

A major conversion was planned for 5th February, entirely confined to RMCs, when routes 716/A and 718 were due to receive RPs, the displaced RMCs being intended for routes 330/A (worked by St Albans and Hemel Hempstead) and 499 (Dartford) with the remainder scheduled for Grays to continue the conversion of route 370. Insufficient delivery of RPs led to a late change of plan and only route 718 was dealt with on 5th February. Harlow lost all its RMCs but Windsor retained a pair for peak journeys on routes 445 and 460. Dartford was the main recipient but Grays received two, which were joined by four more on 11th March which was the deferred date for the 716/A programme. Addlestone and Stevenage lost all their RMCs to St Albans, Hemel Hempstead and Grays but Hatfield retained some which were then downgraded as buses for the 341/B. Other minor RMC operations resulting from these changes included journeys on routes 401, 423 A/B/C and 480 at Dartford; 301, 302, 312, 316A, 318, 334/A/B, 341, 377/A/B and 378 at Hemel Hempstead; and 341 and 343 at St Albans.

With no respite in RP deliveries, it was the turn on 25th March for routes 704 and 705 to lose their double deckers, Dunton Green and Windsor being cleared of RCLs, some of which moved to Crawley and Reigate to take up work on route 405/B (with workings at Crawley on 426A, 434 and 476/A, and at Reigate on 424 and 430). Too many RCLs were rendered surplus for route 405 so a few went to Grays releasing still more RTs from routes 300, 328, 370, etc.

Only one more conversion to RP remained to be carried out; on 29th April route 715, which had introduced RMCs to Green Line work back in April 1962, would lose them. With regular double deck Green Line operation now nearly ended, the opportunity was taken on 5th April to remove RMC 1464 from its outstation at Riverside whence it was transferred to St Albans for normal bus service, its place as substitution coach being taken by a much older vehicle, RF 164. On 29th April, the RMC allocation at Grays for route 370 was finally completed and Swanley received a number for route 477 (plus journeys on routes 401A, 423/A and 493). Guildford lost its entire RMC fleet but Hertford retained a few to bolster the RMC operation on route 341 and also to convert route 395/A and a single journey on each of routes 331/A and 350/A. This was a very big day for Hertford which, in addition to its new RPs, was also the first recipient of the AN class of Atlanteans for route 310/A. These rear-engined machines quickly proved very troublesome in comparison with the RTs which they replaced and, as a result, RMCs and RTs could frequently be found running as substitutes. This last stage of the RMC coach to bus conversion pro-

gramme could not quite be fulfilled with vehicles released from route 715 so Windsor RMCs were also despatched to Swanley and their workings reverted to RT operation.

The 29th April conversion found all RMCs reclassified as buses with one exception; RMC 1516 had been delicensed since 24th January after losing its roof to a low railway bridge in Hatfield and spent the next three months or so in Aldenham having a new top fitted. This was a standard RM roof without luggage racks which made the vehicle unique among the RMCs; fluorescent lighting of the DMS type was installed. RMC 1516 was not officially reclassified as a bus until relicensed for service at Hertford on 4th May. In the case of the RCL class, all except three had been reclassified as buses by 25th March. The three exceptions were RCL 2226, 2237 and 2250 allocated to Godstone for peak hour and Sunday route 709 which was the only Green Line operation to retain scheduled double deckers. However, as in the past, double deck duplicates, particularly at Bank Holidays, and breakdown substitutions continued to occur and both classes could be found back in Green Line service in this temporary capacity from time to time.

Above **London Country's 'flying Polo mint' symbol was a brave attempt at image building which, though destined to be short-lived by edict of the National Bus Company, appeared on many units of the Routemaster fleet. RCL 2218 was downgraded from coach to bus status on 1st January 1972 when it was transferred from Romford to Reigate for operation on route 414. Colin Stannard**

The new light green and yellow livery of 1972 made a change from the more sombre Lincoln green and cream of former times. The new style fleet numbers made a tangible change from London Transport style but in this case the effect is marred by combining them with taller, Johnston type letters. Godstone's RML 2320 works hospital service 482 shortly after its May 1972 overhaul. Colin Stannard

Below Harlow's RMLs played a major role in the provision of local services within the New Town until the October 1972 arrival of Atlanteans. Just before the changeover this Sunday morning scene, with most of the fleet at home, typifies London Country in its early days with RTs and RFs also in profusion. RMLs 2449 and 2450, which head two rows of Routemasters, have both been at Harlow from new. Ken Blacker

RMLs were also on the move in February 1972 when, on the 19th, Godstone's route 410 was taken over, with the exception of a few journeys which remained RML worked, by the new AF class of Daimler Fleetlines which were a diverted order carrying crisp-looking Northern Counties bodies. Ten RMLs left Godstone for Windsor and a new life on the 407/A, 441 and 457/A with the customary sorties elsewhere including routes 335 and 353. Later, on 19th September, one of these RMLs was transferred, at about one day's notice, to High Wycombe where it was urgently needed on new schools service 372 whose RF had proved too small; the RML lasted officially until 27th

November when an MBS became available although it was probably superseded a month or so earlier than this. Meanwhile, another garage to require a vehicle for urgent augmentation had been Northfleet which had received an RCL, its first, on 13th July although it departed on the 29th for Crawley as one of three RCLs needed for augmentation on route 405. The other two, from Grays, were replaced by RMCs, one each from Swanley and Dartford where their regular operation as special duplicates had ceased. On 14th October, Staines became the sixth 'new' garage to receive Routemasters in 1972 when all of Harlow's allocation was expelled upon receipt

of new Atlanteans for coin-box operation. Staines used their RMLs on route 441 (with journeys on 441C/D, 444, 466 and 469) and Windsor also received more RMLs, completing the conversion of routes 407/A, 417, 441/B and 457/A.

Following upon completion of the RCL overhauls in December 1971, an intensive programme was commenced on the RMLs, lasting right through 1972 and up to June 1973. The great majority were dealt with at Aldenham on the same basis as the RCLs, retaining original body and running units, although 23 were dealt with at garages, mostly Garston but penny numbers at Godstone and Hertford.

Most of those dealt with in-house were repainted and brought mechanically up to Certificate of Fitness standard although three Garston buses, RMLs 2423, 2424 and 2437 were not repainted, as their external condition later testified. The first eight overhauls were in Lincoln green but the emergence from Aldenham of RML 2310 on 12th April introduced a lighter and very pleasant shade which had already made its debut on Godstone's new Fleetlines and was about to become quickly widespread with the influx of Atlanteans. With yellow fleetnames and numbers (the latter no longer in Johnston style) and the new yellow relief band, a very satisfactory livery had been achieved but it was not to last. Quite soon afterwards, the National Bus Company announced its infamous corporate identity policy under which all company liveries, no matter how long established or how well regarded, were to be submerged into a common style strictly dictated from head office at New Street Square. In the case of double deckers, companies with a green livery had to adopt leaf green with a single white band between decks, grey wheels and fleet numbers, white block capital fleet names five inches high placed forward on the vehicle, and a white double-N symbol ten inches high. On 16th October 1972 the first corporate advertising campaign commenced and, on 25th of the same month, RML 2342 was despatched from Aldenham to Northfleet as the last vehicle overhauled in the company's own livery. The first in corporate style, RML 2333, had already been received at Godstone on the 20th and this livery quickly spread. Twenty-seven RMLs had received the light green and yellow style during the six months of its application. The NBC's stated intention in mid-1972 had been that all vehicles of all subsidiaries should be in corporate colours within two

years and, indeed, many companies complied but London Country, beset with problems, failed to meet this target and some of the Routemasters which it was still running up to the last year of the nineteen-seventies were still in pre-corporate colours.

A country-wide craze too strong for London Country to resist was for all-over advertisements and, in October 1972, RMC 1516 was painted by Wright Signs of Cockfosters in white with intricate multi-coloured design and lettering advertising Welwyn Department Store. It was allocated to Hatfield for route 341/B. This was the vehicle which, earlier in the year, had been reroofed; in January 1973, a severe rear end collision required further rebuilding from which it emerged in February with a revised advertising design on the replacement panels. In March, RMC 1490 became the second all-over advertisement when the same contractor that had dealt with

Windsor's RML 2458 was one of many received from Harlow for the Routemastering of the 441 group in October 1972. Repainted in National Bus green and white in March 1973, it still carries the nearside siting post fitted to several green members of the class as an aid to drivers' nearside visibility, which some retained throughout their time with London Country. S. Clennell

Overall advertising designs are by their nature normally gaudy and the Welwyn Department Store livery on RMC 1516 was no exception, but at least it brightened a very rainy day in Hatfield. An interesting feature was that the old Green Line window mouldings were picked out as part of the overall design. Ken Blacker

RMC 1516 again applied a predominantly white scheme, this time rather less embellished, as a mobile hoarding for London and Manchester Assurance, a guise which it was to bear for four years although the design underwent change. First allocated to Grays, it led a nomadic existence travelling annually from Grays to Reigate, Northfleet, Windsor, High Wycombe and Garston, always in that order and spending an average of two months at each. The circuit finally ceased when the vehicle was delicensed on 1st April 1977 for repainting into standard livery. Except for Grays, the garages which it routinely visited were not normal users of RMCs. RMC 1516, bearing an advertisement of only local interest, remained at Hatfield and continued to do so for several months after May 1974 when it adopted a new, mainly blue and orange guise advertising Fine Fare, who now owned the Welwyn Department Store. Its final few months were spent based at St Albans operating on route 330 which actually passed the shop whereas route 341/B did not even serve the town. Concluding the subject of London Country's advertisement Routemasters, RMC 1480 became the third, in March 1974, again carrying a basically white scheme, this time in favour of the Invicta Co-op which it advertised on Dartford's route 499 through to August 1975.

24th February 1973 saw the first and only major Routemaster route reallocation of the year consequent upon ANs striking into Grays for the 328/A/B which were renumbered 328/329/373; however RMC and RCL journeys remained in the peak hours on all three routes. Surplus RMCs moved out to commence taking over part of route 406. Reigate received its full complement but only two were available for Leatherhead and even their hold was tenuous; a phased recertification and repainting programme was planned to commence in May, leaving little scope for increasing Leatherhead's quota. Timetable alterations on 7th July saw the replacement of Windsor's route 417 with short workings on route 458 and the renumbering of 457A/D as 452, whilst at Northfleet a new route number for the RMLs was 488, a renumbering upon diversion of part of the long established 487.

These changes apart, the year was one of remarkable theoretical stability but, under the surface, all was far from well. Huge trading losses in its formative years had marked London Country as the greatest problem confronting the NBC; now its position was deteriorating further due to a growing crisis over vehicle spare parts which was hitting the country as a whole but some operators more than others. On London Country, the problem

Above **London & Manchester's RMC 1490 had large areas of plain white as part of its paint scheme. Seen in Redhill with home made destination blind, it was on its third and final stint at Reigate when photographed in July 1975.**
Paul Hulyer

Right **Another mainly white paint scheme was applied to RMC 1480 for Invicta Co-op, seen in Dartford.** Steve Fennell

had begun to get desperate by the latter part of 1973, and it was highlighted by the delicensing at the end of the year of 22 long-term unfit Routemasters. Although there was every intention of returning them to serviceability as soon as possible, the inevitable cannibalisation which occurred only served to make the position more precarious and, for some, this was the end of the road. Unbeknown to everyone at the time, RMLs 2306 (at East Grinstead). 2319 (Godstone) and 2426 (Northfleet) had come to the end of their lives. The East Grinstead vehicle was of interest in that it was never replaced and, even though an RML was officially scheduled at this garage until July 1977, no vehicle was ever available to replace it and the working had to be covered by borrowing from Godstone. 1974 proved to be a crisis year during which a further six RMLs breathed their last and the first involuntary withdrawals occurred in both of the other classes with the delicensing of Hatfield's RMC 1509 in March and Reigate's RCL 2227 in October. 1975 showed no respite; six more RMLs and a further RCL were permanently delicensed. A chronic engine shortage prevailed but many specialist Routemaster parts were also in short supply. London Transport, the only source for some of them, was itself running into severe trouble and frequently unable to help.

Impending Certificate of Fitness expiries brought about a recertification programme for the RMCs which got under way in May 1973 with accident damaged RMC 1484, the first of its class to appear in National green, and began to gain momentum from July onwards. Some of the recertifications were carried out in garages, principally Romford and Garston, but 48 vehicles received Aldenham overhauls without any swapping of bodies. The last to be overhauled at Aldenham under this programme was RMC 1455, completed in May 1975, although one further London Country Routemaster was dealt with there in the form of RMC 4 whose third and final overhaul lasted from May 1975 right through to June 1976. It ended with the vehicle returning to Hatfield in October after a spell in the Chiswick experimental shop and now sporting opening front windows. The RMC class was the only one of the three to go over completely to NBC green although the corporate image was not 100% achieved as RMCs 1468, 1497, 1500, 1501, 1507 and 1512 received non-standard grey central relief bands instead of the stipulated white. Even long term delicensed RMC 1509, never to run again, was repainted in the vain hope that it would one day be rendered mobile. The same thing happened

Top **RMCs arrived on route 406 in February 1973, almost sixteen years after the brief trial operation here of RM 2. RMC 1461 was one of the vehicles received at Reigate for this purpose after an influx of Atlanteans had rendered them surplus to requirements at Grays.** Colin Stannard

Centre **The RMC recertification programme which commenced in May 1973 found the whole class in NBC livery in due course. Vehicle and spare parts shortages felt from the end of that year were intense by 1975 when Hemel Hempstead's RMC 1495 was photographed on loan to Garston covering for an RML on route 306.** Capital Transport

Right **RMCs made occasional returns to Green Line work for special events such as racing at Brands Hatch. RMC 1493 is seen at Lewisham in April 1976.** Paul Hulyer

RCL 2237 received its unique National Bus livery with Green Line fleetnames at a Leatherhead garage repaint in March 1975. It served for a year as spare vehicle for route 709 but was allocated for most of the time to Crawley and could normally be found at work on route 405. J.H. Blake

with RCLs 2225/2227. A repainting programme began on the RCLs in 1974, taking advantage of the fact that many were unfit for service and could therefore be repainted without causing lost mileage. Many were towed, sometimes engineless, to Leatherhead where the first batch of repaints was carried out, and later to Grays. The first corporate coloured RCL, 2244, was in fact in the engineless category and the first to run in service in the new colours was RCL 2256 at Crawley in June 1974. In March 1975, one of the three RCLs still classified as a coach, RCL 2237 which spent much of its time temporarily allocated away from Godstone to Crawley, received NBC livery complete with corporate style GREEN LINE fleetname and was unique in doing so. The other two Green Line vehicles plus eight others remained in their old colours to the end.

From the service revision point of view, 1974 was fairly quiet on the Routemaster front. Conversions to omo had abated but area revisions lay ahead for Welwyn Garden City and Hertford which found RMCs bearing new route numbers. 16th February saw the withdrawal of many of Hatfield's traditional RMC operations and their replacement by routes 840/842-846 (and later 848). On the same date, Grays route 300 was renumbered 375 so that route 303A could be renumbered 300. At Hertford, on 4th May, route 395A was renumbered 395, and routes 331A and 350A which had one RMC journey each were renumbered 337 and 351 respectively. Route numbers bearing suffixes were being eliminated to accommodate the spread of Leyland Nationals with three-track route number displays under a process which had begun in 1973, other routes with Routemaster journeys affected having included, up to 4th May 1974, 316A (renumbered 322), 328A (329), 328B (373), 330A (330), 334A/B (334), 341B (341), 377A/B (377), 457A/D (452). On 22nd June, reductions

at Windsor and Staines on route 441 (also journeys on routes 335 and 353), theoretically released RMLs to Hemel Hempstead for 334, 377 and 378 and numerous works journeys to Apsley Mills where extra capacity was needed. However, the unstable vehicle position generally meant that RMCs predominated on these workings.

By 1975, the acute shortage of serviceable buses meant that the official allocation of vehicle types to routes often could not be achieved and unusual workings almost became the norm. Standard omo types often appeared in place of Routemasters and sometimes this happened in reverse. To avoid total collapse of some services, buses were hired in from a number of other operators, commencing in 1974 with Merlins from London Transport and followed in 1975 by Bristol MWs from Royal Blue, Daimler Fleetlines and Roadliners from Bournemouth, Leyland Titans from Maidstone and Southend, and AEC Regents from Eastbourne. With almost half the RML fleet out of commission by midsummer, such desperate measures gave the only breathing space. In July, the purchase of three of Southdown's handsome full-fronted Leyland Titan PD3s provided much needed assistance to Godstone's RMLs on routes 409 and 411 on which they ran as the LS class still in Southdown's traditional green and cream colours but with NBC style LONDON COUNTRY fleet names. The worst hit garage, where maintenance standards and staff morale appeared to be shambolic, was Garston which had scraped up RMCs on loan to help out, mostly from Hemel Hempstead and Grays but, even so, RTs far outnumbered Routemasters on scheduled RML workings. On 31st May, Garston received, though only on paper, ten more RMLs to replace RT journeys on routes 301/B, 302, 318 and Wednesdays 345. Conversion on this date to one man operation of Hemel Hempstead's nominally RML worked

334 and 377, and the withdrawal of 378, reduced its Routemaster requirement although new crew work was gained on routes 301 and 302. Only two RMCs were lost but a first time recipient of the class was Tring, for a newly gained allocation on route 312 plus journeys on route 387. On 30th August, the Tring operations were officially converted to RML but, although one RMC was replaced by an RML later in the year, the second of the class never arrived. Hemel Hempstead's crew workings on routes 301, 302 and 334 were also officially converted from RMC to RML on the same day.

On 30th August 1975, renumberings in the Dartford area affecting RMC workings saw routes 423A/B/C becoming 423 and 499 whilst at Hatfield some of the RMC journeys introduced in February 1974 were discontinued on the same date (routes 843-846) or earlier (848). A theoretical change was the introduction of Leatherhead RMC journeys on route 408 but this garage had no Routemasters serviceable at the time. Relief finally came in November when, starting officially on the 8th but in practice just afterwards, Leyland PD2s from Maidstone Borough Transport took over from RMCs on Dartford's 499 (plus 423 and 480 journeys), releasing six RMCs to help the hard pressed fleet at Garston. From 5th January 1976, Swanley's route 477 was partly turned over to AEC Regents hired from Eastbourne, but only one serviceable RMC was available to transfer elsewhere and this went to help out a bad RML position at Northfleet. Northfleet had been a fairly consistent borrower of RMCs, as had Harlow and even Stevenage which, at this stage, did not have any Routemaster allocation at all but found work for them. January 1976 saw the removal of Hemel Hempstead's last RMLs to locations which had more urgent need, RMCs covering despite their lower capacity until replacements were available towards the year's end.

An increase in omo at Grays on 3rd April 1976 saw the loss of two Routemasters although new route 324, replacing 370A, was introduced with scheduled RMC and RCL journeys. One of the displaced RMCs travelled south to Chelsham to begin a long drawn out replacement of RTs from route 403 and their many journeys on route 453. Next month, on 15th May, route 709 finally succumbed to one-man operated Leyland Nationals and the last of the Green Line RCLs were reclassified as buses though RCL 2237 retained its coach fleet name for the remainder of its time with the company. At the same time, reductions took place at Hemel Hempstead and Hertford. This was a direct result of NBC policy whereby its subsidiaries had been instructed in 1975 to seek substantial subsidies to maintain services which, if not forthcoming, would result in reductions or even total withdrawals. This meant, for instance, that Hertfordshire County Council would be asked for almost £1.5 million against the existing subsidy of £470,000, a demand which it could not meet. A similar problem existed in Surrey. The Hertfordshire reductions made RMCs available, which joined by RMCs released from Grays by the intake of surplus RCLs from Surrey cuts, furthered the conversion of route 403 and also enabled a firmer Routemaster hold to be made at Leatherhead. A lone RCL which had been at Northfleet since March covering RML recertifications was removed at the same time. After service on 12th June, the Eastbourne Regents went home in time for their summer seaside season whilst, on 4th September, the Maidstone PD2s followed suit. One-manning

The protracted conversion of route 403 to Routemaster was assisted in July 1977 by the conversion to SM of Dartford's 499. Corporate-liveried RMC 1481 moved across to Chelsham on this occasion. Peter Plummer

Two of the three RCLs which enjoyed a four-year stay of execution in the downgrading from coach to bus status retained traditional colours. Representing the last vestige of the old Green Line era, RCL 2226 purrs towards its home base, Godstone. Peter Plummer

at Garston on routes 306 and 311 enabled the six RMCs to be returned to Dartford. Many of the Watford services bearing suffix letters were renumbered into the 800 series at this time, as was Napsbury Hospital service 345 which, though officially RML was actually RT worked; it became 834. After strenuous efforts all round, the vehicle position was, at last, showing signs of improvement as witnessed by the withdrawal, also on 4th September, of the ex-Southdown PD3s at Godstone. 30th October 1976 found RMLs in former Maidstone & District territory when Northfleet's route 480 was extended at Denton into the Valley Drive area. An offshoot of the 480, numbered 482, which had come into being on 30th August 1975, was now withdrawn, being covered by revisions to the 487/488.

Hemel Hempstead saw the first round of Routemaster activity in 1977 when local services were renumbered into the H series on 8th January. RMCs now covered odd journeys on routes H13 and H15, the latter having a St Albans involvement being worked off route 330. 29th January saw wide scale one manning at Windsor which removed RMLs from all except routes 407/A, the 441/B and 458, enabling eight of the class to travel eastwards to re-establish themselves at Harlow. The venerable RT class, which had so often come to the rescue when Routemaster and other types had failed, was now itself experiencing serious shortfalls as vehicles reached the end of their economic lives and at Harlow a batch of South-end Leyland Titan PD3s had been covering their duties. These now returned home to leave routes 339, 396, 397/A and a host of journeys on town services such as 804, 805, 807, 810A and 811 in the hands of RMLs. The same RT shortage resulted in a new hiring from 4th March when Maidstone Atlanteans joined the RTs and RMCs on Chelsham's route 403. 2nd April saw the closure of Tring garage and the withdrawal of its main Routemaster route 312 and, on the same day, RML work at Harlow contracted considerably to just routes 339 and 807 with a few new journeys on route 810. 7th May saw the demise of RMC/RCL workings at Grays on routes 324 and 375, whilst on 21st Northfleet's 487 and 488 went over to MBS operation. The first half of what was to be a busy year for Routemasters ended on 1st June when the bulk of the new BT class went into service at Grays. These Bristol VRs, allocated by NBC headquarters with its usual scant regard for its subsidiaries' preferences or standardisation plans, were placed into service ahead of a big omo scheme at Grays, permitting RCLs to make their debut at Chelsham on route 403 and enabling the full conversion of route 406 to RMC to be achieved at long last.

Implementation of the one-manning at Grays on 9th July left just two RCLs running a few journeys on routes 323, 328 and 371, and saw a further exodus of RCLs including a pair which went to assist the RMCs at Dartford, the first RCLs at this location. Godstone and East Grinstead's routes 409 went over to Atlanteans on the same day, although a few Godstone worked RML journeys remained. On the 16th, Harlow lost the remainder of its scheduled RML workings to SNB operation although two RMLs remained. They were often to be found on new Green Line routes 702 and 703 and also served as frequent performers on town route 808. 23rd July saw the

RCL 2244 was a regular performer at Dorking from January 1972 onwards and was the last Routemaster at the garage when withdrawn from service in June 1978. After Dorking's last scheduled RCL workings were lost with the one manning of route 414 in October 1977, RCL 2244 stayed on as a semi-regular performer on the 449. Peter Plummer

conversion of Dartford's 499 to SM omo working, some RMCs passing to Chelsham for route 403, some going to help out at North-fleet, and yet others remaining as unscheduled spares at Dartford. All semblance of order was now beginning to vanish as Hertford received two RMLs in July, which could normally be found on Atlantean routes 310 and 316. This followed the allocation in late April of a pair to Hatfield, also its first, whilst an odd one was also to be found for a few months at Stevenage, principally on route 303/C, and another at Dorking bearing makeshift front blind displays for route 414.

High Wycombe garage closed on 1st October 1977. The year had been one of major contraction with no fewer than four garages closing, High Wycombe being the fourth. Tring has already been mentioned; Luton had shut on 29th January without ever having a Routemaster allocation and Romford, one-time home for so many RCLs in their prime, ended its transport operations on 2nd July. The last Routemaster to be repainted in London Country ownership was dealt with during the year, the honour falling to RCL 2248 of Grays which appeared in NBC colours in February. London Country was committed for economy's sake to adopt total one-man operation as soon as possible and the Routemasters now presented themselves as a stumbling block to this progress. In October, after nearly a year of rumour, 38 of them, all non-runners, were officially put up for sale with the first approach being to London Transport in the logical expectation that they would be the most likely buyer. The bulk of the 38, many heavily cannibalised and several engineless, consisted of RMLs of which there were 29, together with 7 RMCs and 2 RCLs. A deal

was concluded, but London Transport also urgently needed runners for immediate operation and, in December, agreement was reached for thirty serviceable vehicles to change hands as well as the 38 non-runners. Many of the dead vehicles had been towed to Grays and were looking very forlorn in the garage yard.

The High Wycombe closure coincided with the one-manning of its services from their new base at Amersham, the RMLs going as a group to Garston to make up its serviceable complement at long last. Routes 441/B/C from Staines and Windsor also succumbed to Nationals although one or two RMLs remained temporarily to cover shortages. In the south, route 414 received SNBs except for a few remaining Reigate based crew journeys, but Dorking retained an unofficial RCL as a semi-regular performer on route 449 and sometimes even on Green Line 714. This enabled the Maidstone Atlanteans on route 403 to be returned from loan. Hatfield, Hertford and St Albans were all hit by the arrival of SNBs on routes 330, 341 and 395 on 19th October but a few RMCs remained as unscheduled spares including RMC 4 at Hatfield. A week later, a similar fate befell route 405, marking the end of the RCLs at Reigate although a few remained at Crawley for town service 405B. Massive inroads by Leyland Nationals during 1977 had actually brought about a surplus of serviceable Routemasters by the end of the year, some being transferred to training duties to cover for defective LR class ex-Ribble PD3s. On 22nd December, the first three of the sold vehicles (RMCs 1489, 1491, 1494) returned to London Transport stock with many more following before the year was through.

1978 was to prove a watershed year for the London Country Routemasters. It began with 133 of the original 210 still licensed for passenger service (45 RMC, 33 RCL, 55 RML) and ended with this number almost halved. The company's troubles remained acute with the unreliability of many Atlanteans and, to a lesser extent, Nationals giving a problem which was only compounded by the failure of Tinsley Green works (which had opened with a flourish as recently as January 1976) to maintain a satisfactory output, resulting in the recertification programme for Swifts falling ever more behind schedule. At the other end of the vehicle age scale, the venerable RTs, whose last official foothold was at Chelsham on route 403, were almost finished. Routemasters were called upon daily to cover deficiencies in other classes, but their ability to do so was to be curtailed from early February onwards with the expiry during the following twelve months of the Certificates of Fitness for the whole RCL class.

At the start of 1978, all three Routemaster classes could be found somewhere or other in the fleet deputising for omo vehicles. This was made possible by the availability of surplus conductors at a number of garages, even those officially devoid of crew work, through the absence of an agreement for compulsory redundancy. Thus, although RMCs were officially scheduled only at Reigate and Leatherhead for route 406 (with journeys on route 418) and Swanley for route 477 (plus 423 and 493), examples could also be found at Hatfield on routes 341 and 840 group, Hertford on routes 310, 316 and 395, Windsor on the 446 group, Dartford on route 499, Grays generally working alongside RCLs, and at Hemel Hempstead covering in the absence of RMLs. Likewise with the RCLs whose official allocation was to only three garages: Chelsham for route 403 (with 453 journeys), Crawley for route 405B (405, 434) and Grays for a miscellany of workings on routes 323, 328 and 371 plus contracts. Unscheduled RCL workings were to be found at Dorking mainly on route 449 but also on route 425, Dartford alongside the unscheduled RMC on 499, and Godstone where several were deputising for AFs, ANs and RMLs on routes 409, 410 and 411. Even the RML class, which had been reduced more heavily by withdrawals than the other two, carried its share of unscheduled work. Official allocations still remained at Garston and Hemel Hempstead for route 347/A (plus 311, 318, 833, 835) although in practice Hemel Hempstead's were all RMCs, Godstone and Reigate for route 411 (plus 409 and 410), Windsor for route 407/A (with 458) and Northfleet for route 480. Windsor's fleet was much augmented by additional RMLs which could usually be found on routes

441, 452, 457 or 458 whilst Harlow kept a few at work, usually on Green Line 702 and 703 but also occasionally on town services. Finally, there were a couple at Hertford keeping the unscheduled RMCs company. Further variety was added in January 1978 when Addlestone borrowed an RCL which was kept busy on routes 420, 427, 437 and 461 during peak periods for three months. A few vehicles were now beginning to sport a revised NBC double-N logo in red and blue on a white square, but many more were starting to acquire the uncared-for look which was to characterise London Country's Routemasters in their later days.

The steady pace of RCL withdrawals resulting from 'ticket' expiries saw a reversal of roles at Chelsham commencing early in the new year with RMCs replacing RCLs, the first vehicles received being from unscheduled sources elsewhere, such as Hatfield, Grays (where RMC operation ceased for a few months) and Dartford (where it ceased entirely). March saw an easing of the vehicle position with the delivery of the first of 79

single-door Atlanteans ordered for 1978; these quickly went into service at Crawley where, by 5th April, the official omo conversion date of route 405B, Routemaster operation had become very spasmodic, most RCLs already having gone to ease shortages at Chelsham and Godstone. Hertford's last Routemasters also departed in April, the RMLs being withdrawn and the RMCs being despatched for service elsewhere including one, surprisingly, to Stevenage where it remained for much of the year. RMC Certificate of Fitness expiries resumed in June, witnessing the withdrawal for this reason of more than half a dozen of Hemel Hempstead's contingent though official conversion of their operations to omo did not come until 2nd September and two struggled through until then. On 20th May, local services in Welwyn Garden City and Hatfield were again renumbered, this time into the G series, and the two RMCs which still remained at Hatfield (4 and 1512) were often to be found on these, although journeys were sometimes made on route 722 (the successor to the northern end of route 716) and later the tendency

was to work on route 330. The 2nd September conversion of routes 347 and 347A, which included the renumbering of the latter to 348, came ahead of the full delivery of Leyland Atlanteans, and resulted in a few RMLs remaining at Garston for a few additional weeks, even being provided with 348 blinds. Two further conversions remained in the 1978 programme; routes 406 and 411 were both scheduled for omo conversion on 28th October. In the case of route 406, Reigate and Leatherhead both began to lose RMCs well in advance as Leyland Nationals were drafted in from July onwards, whilst route 411 started receiving Atlanteans a month ahead of the designated date. Reigate's three RMLs all found a new use as trainer buses (although one was later returned to service elsewhere), as did two of Godstone's. By October, Godstone retained only two RMLs and two RCLs of its once large Routemaster allocation and even the use of these was irregular. However, contrary to most expectations, one of the RMLs was retained and continued in service at Godstone as a spare vehicle until well into 1979.

By October 1978, route 403 was almost entirely back to RMCs from RCLs and Grays was re-establishing its RMC stock as the RCLs faded away. At the year's end, only three RCLs (2237, 2249 and 2250) remained in service, all at Grays. In addition to their normal miscellany of local stage and contract work, it was not uncommon for them to appear on Green Line route 723 in a nostalgic reminder of the role that these splendid vehicles played in their heyday. Particularly poignant, when it appeared, was RCL 2250 which wore its original Lincoln green livery to the end.

Month by month throughout 1979, Certificate of Fitness expiries made heavy inroads into the surviving Routemaster fleet and, of those which were theoretically operational, many were, in fact, unfit for long periods. The end of the RCLs came in January and it is thought that RCL 2250 (by then the only one with a surviving CoF) made the last run on a schools special at Grays on 24th January. Thereafter, a few RMCs remained at Grays even though, from 6th January, their scheduled work on routes 323, 328 and 371 had

ceased, but their numbers declined as certificates expired and all had gone by October. All of the other unscheduled allocations had ceased before this as Hatfield, Staines, Godstone and finally Harlow lost their remaining Routemasters. RMLs from the last three moved to Northfleet where a whole series of certificate expiries throughout the summer produced a grave situation on route 480, on which a whole variety of newer, mainly single deck types, were giving assistance when available. Northfleet's stock of RMLs, many of which had not been repainted for several years, looked unbelievably down at heel and it was clear from a glance that the end was drawing nigh. Only one omo conversion affecting Routemasters took place in the early part of the year; this was the replacement at Chelsham by Atlanteans on 3rd March (deferred from 27th January) of RMCs from the main 403 service, although the Routemasters still remained on the express part of the 403 and put in even more appearances on the 453 to utilise crews for the full day, the 403 Express being only a peak hour operation.

On 1st May, RMC 4, the last of its class at Hatfield, was withdrawn from regular service after a career which had extended far beyond those of the other three prototypes. However, its future was secure. Despite having designs on the rest of the London Country Routemaster fleet, London Transport did not wish to purchase such a non-standard vehicle and, in any event, its owners had decided to retain it as a display bus and it was sent to Hertford garage for a repaint. One of the first functions in its new role was to duplicate the final journey on the last day of operation of Green Line route 709 on 26th October 1979. Meanwhile, negotiations for the sale of the remaining fleet to London Transport reached a successful conclusion but not until after a disagreement between the two parties over the condition in which vehicles should be handed over resulted in LCBS selling fifteen Routemasters direct to dealers Wombwell Diesels in March. Stung into action, London Transport quickly came to an agreement with Wombwell but too late to save RMLs 2423 and 2424 from being broken up. The remainder, with London Country's agreement, stayed on until London Transport could take them. With large numbers of withdrawn buses (not just Routemasters) in stock, London Country was experiencing severe pressure on its parking space which culminated in the short term leasing from Adams Foods of a large open area in Colney Lane, on the old Radlett aerodrome site, upon which withdrawn buses began accumulating in April 1979.

With the approach of winter, the final stages in the saga began taking place. From 17th November, ANs started to take over on route 480, a process which was complete by 19th December. They came just in time, the rolling stock shortage being so serious that, quite unexpectedly, two training buses were put back into service at the end of October, these being RML 2446 and RMC 1500. On 17th December, another of the remaining Routemaster operations began to give way with the receipt of the first Atlanteans for Swanley's 477. However, this did not prevent the surprise transfer into Swanley just before Christmas of its first and only RML, 2445, nor of RML 2452 to Chelsham, both recently released from route 480. In these final stages, almost anything could happen and, in January, Swanley's lone RML, 2445, passed to Chelsham where it is thought to have run until 13th February. Atlanteans invaded Chelsham's 403 Express in January 1980 and finally, on 16th February 1980, Windsor's 407/A succumbed. This officially marked the end of open platform buses on London Country with RML 2422 withdrawn on this date from

Top RMC 4 was withdrawn from regular service on 1st May 1979 after leading a remarkably long and active life for a very non-standard prototype. Hertford garage made a fairly respectable job of repainting it, and the vehicle is seen later in the same year at London Transport's Shillibeer celebrations. The Leyland rear hub cover is a new innovation for a Routemaster. Capital Transport

Centre and Right Some of Northfleet's RMLs were remarkably down at heel towards the end of the type's responsibilities on the 480. RML 2422, a comparatively recent, August 1979 arrival from Garston in Lincoln green with white band, looks quite smart compared with RML 2343, a Northfleet bus for the whole of its country career. Both were photographed at Gravesend in late-summer 1979. Peter Plummer/Colin Brown

service at Windsor along with RML 2446 which had been retained temporarily at Northfleet. The Windsor bus was noteworthy in wearing Lincoln green livery right to the end, albeit with NBC lettering applied during recertification at Garston in 1976. With a commendable sense of history, the company arranged a special farewell tour from West Croydon on Saturday 1st March, covering routes 403, 477 and 480 with RMC 1512 and RML 2446, the latter having been partially repainted for the event. It was anticipated that this would be the end but, on 4th and 5th March, Swanley unexpectedly put RMC 1512 out on route 477 in the absence of an available Atlantean. This was positively the end for, next day, all remaining Routemasters, including some which had served latterly as trainers, passed into London Transport ownership.

An interesting feature of the RML class remains to be mentioned; this is the unusual stability which a high proportion achieved during their time with London Transport and throughout the London Country regime. Of the one hundred buses, no fewer than 43 spent their entire working life at the garage of original allocation, including 15 at Garston (RML 2420, 2421, 2423–2425, 2427, 2429-2435, 2437, 2438), 14 at Northfleet (RML 2323-2328, 2337-2343, 2345), and 10 at Godstone (RML 2312, 2314, 2316, 2317, 2319, 2330-2332, 2334, 2335). Garages with a single 'long stayer' each were East Grinstead (RML 2306), Reigate (RML 2308), High Wycombe (RML 2417) and Windsor (RML 2436).

Left with only RMC 4, London Country took the opportunity to repaint it once again, this time back into original Green Line livery complete with raised bullseyes. The inspiration for this was the Green Line 50th anniversary celebrations scheduled for July 1980 but its debut in this new guise was made at Cobham Museum open day on 13th April followed by an appearance at Battersea Park on 3rd May. After being located at a number of garages, in March 1981 RMC 4 reached Dorking which was its home until the autumn of 1989 when it moved to Crawley. With the splitting up of the London Country organisation on 7th September 1986, it passed to London Country Bus (South West) Ltd, moving out of public ownership with the acquisition of South West by Endless Holdings (Drawlane group) and Speyhawk Land & Estates Ltd upon privatisation on 22nd February 1988. Though still retained by London & Country principally for display purposes, RMC 4 has been known to put in an occasional appearance in public service as a last resort in the absence of other rolling stock.

Top **Though now primarily a display bus RMC 4 saw occasional service as on the last day of Green Line route 709, 26th October 1979, when it performed the final run.** D.W.K. Jones

Centre **With a couple of ticket rolls as the only decoration and with misspelled blinds, RMC 1512 and RML 2446 stand at Northfleet garage prior to London Country's Routemaster Farewell Tour on Saturday 1st March 1980.** Eamonn Kentell

Left **RMC 4 was restored to traditional Green Line livery, including restoration of the raised bullseyes, in time to make its debut at the Cobham bus museum open day on 13th April 1980. It continued to make occasional special trips in public service, as in this view at Queensbury on 20th June 1981.** Mike Harris

CHAPTER THREE
THE FRM

The last week of June 1970 had found FRM 1 in Aldenham undergoing its first repaint, from which it emerged with the same mixture of gold fleet names and white and black bullseyes front and rear as before, the non-underlined LONDON TRANSPORT having by now been adopted as standard for the Routemaster fleet. A year later, its gold fleet names were replaced by white outline bullseyes of the type used on double deckers from spring 1971. When SMS class single deckers took over route 233 on 27th March 1971, providing an improved frequency at busy times, FRM 1 was shuffled across to route 234, a larger operation than route 233 but still Croydon based and dominated by Atlanteans. This haven disappeared on 20th January 1973 when, as part of a mass onslaught on the Croydon area by new DMS Fleetlines, route 234 was converted as the first step in withdrawing the XA class. Being a single door bus, and therefore incompatible with the DMSs, the FRM ceased operating and was delicensed and placed into storage on 1st February.

Once again, a decision on the future of FRM 1 was required; should it continue in service and, if so, where? No work being readily apparent, it was given a leisurely

overhaul at Chiswick, during which time a new base was identified at Potters Bar where the local circular service 284, like the 233, required only a single bus. The overhaul was completed in July. A single alteration to its physical appearance during overhaul had been caused by the insertion of an air intake in the middle of the front panel where the bullseye transfer had previously been. The livery now included the standard white central band and yellow doors, the latter feature being hailed as a 'dashing change of livery' which would greatly assist passengers by indicating where the entrance was positioned! The latest 17in filled white roundels, proclaimed as another major step forward in design standards, replaced the gold fleet names, and sensibly sized fleet numbers were applied for the first time, again in the latest white version.

FRM 1 was relicensed at Potters Bar for driver training on 29th August 1973 and, on 13th October, took up its new role on route 284. Now at the opposite extremity of the London network from its previous base, the vehicle led an even quieter existence than before as route 284 was seldom busy. It never strayed on to other routes; indeed the running number PB71 was actually painted on.

Above **Transferred to the 234 and 234B routes, and here incorrectly displaying via points for the former, FRM 1 now has more chance to show its paces than it did on the 233. The gold fleetname formerly carried on the sides has now been replaced by a DMS type open white bullseye which better matches the ones on the front and rear.** Gerald Mead

Above Right **At Potters Bar, FRM 1 gave the appearance of being in the hands of a garage which really cared, and was always in a more polished condition than at its two previous locations, even to the extent of its chromium plated front wheel trims. It is seen in Mutton Lane, Potters Bar, during 1975.** Capital Transport

Right **FRM 1 during its last summer in service, photographed at work on the Sightseeing Tour in July 1982 while allocated to Stockwell. When repainted into traditional red and cream livery, a red and gold metal bullseye was added at the front above the air intake.** G.A. Rixon

Despite the limited nature of its work, confining it within a two-mile radius of the garage, FRM 1 was treated with great affection by the staff at Potters Bar and it was even embellished with polished front wheel trims removed from an SMS. However, by 1976, its days in service once again appeared to be numbered. Route 284, which was hardly a money spinner at the best of times, required County Council subsidy to keep it going and this was not forthcoming. However, the FRM did not wait until the end. It rammed into London Country's Leyland National SNB 92 in Mutton Lane, Potters Bar, and was delicensed on 30th September. Its days on stage carriage work had come to an abrupt end.

A change of career lay ahead at Stockwell on the Round London Sightseeing Tour but, before this, the whole of 1977 was spent largely inactive. The accident damage was repaired and an Aldenham repaint followed in November 1977 to the same style as previously; in December an internal public address equipment system was installed ready for tour work. Allocated to Stockwell on 27th January 1978, the vehicle began touring alongside much newer, but far less comfortable, DMSs on about 14th February. Despite odd spells of absence, the longest being from August 1980 to March 1981, FRM 1 came to be regarded as something of a high class flagship for the Round London tour fleet, its exclusivity being acknowledged when repainted in February 1981 into traditional London Transport colours of red and cream with gold fleet numbers and underlined names: a livery which actually predated the FRM but looked good nonetheless. On 29th November 1982 came a reallocation of the entire tour fleet, including FRM 1, from Stockwell to Victoria, but this was the last reallocation this unique vehicle would undertake while in service. A management unsympathetic to it was now in place and had decreed withdrawal, which came on 3rd February 1983 when the FRM was delicensed into Edmonton garage for storage. It was not, however, allowed to fade away and was quickly retrieved and taken to Chiswick. In April 1983 it was handed over to the London Transport Museum for preservation, its future assured, and though not on regular display at Covent Garden it has made occasional public appearances at special events.

With an aggregate total of approximately seven years' stage carriage work followed by more than four years active touring, FRM 1's tally of revenue earning operation was notably high for a prototype. It repaid its cost far more handsomely than many of the subsequent off-the-peg Merlins, Swifts and Fleetlines whose lives, in some instances, were scandalously short. However, because it was overtaken by circumstances, it really contributed little to London bus development and its fame far outweighs the meagre contribution which it made. FRM 1 was the latest, and may remain the last, of a sizeable number of experimental one-offs commissioned by London Transport and its predecessors. Though some, like FRM 1, contributed little in the long term and others were basically unsound in concept anyway, none should be decried for, if nothing else, they were proof of a vibrant, go-ahead organisation not afraid to originate and lead the field.

CHAPTER FOUR
ALL-OVER ADVERTISEMENTS

The nineteen-seventies brought an era of new awareness of the potential of the double decker as an advertising medium unique in its bulk and mobility. As a result the Routemaster has been used in a number of interesting and colourful – if not always tasteful – guises in marked and total contrast to the customary staid appearance of the traditional red London bus. These total breaks from tradition have taken two forms; first was the 'sale' of complete vehicles to advertisers or their agents to paint as all-over advertising hoardings for periods of time specified by contract. Later came London Transport's own specially-devised liveries, usually to commemorate some special event, but with a strong advertising bias to cover the cost and hopefully produce a good profit on the scheme.

The very first all-over advertisement project appeared as a tentative one-off in the final months of the pre-GLC era and created something of a mild sensation at the time. London Transport was by no means the first bus operator to adopt this tactic as a source of revenue but there had been a general expectation that it would stay aloof from the trend. There was therefore great surprise, and in some quarters horror, when RM 1737 appeared on route 11 from Riverside garage in August 1969 in a multi-coloured but predominantly pale green design embellished on the lower panels with guardsmen in full parade regalia including red jackets and bearskins, and carrying the legend between decks 'Silexine Paints welcomes you to London'. To Silexine Paints Ltd had gone credit for breaking the traditional

London Transport reticence, and also for stimulating artistic effort by running a competition with £100 first prize amongst five London art colleges for the best design. It was won by David Tuhill, a final-year student from the Royal College of Art and painted by Walter Hyde of Magnet (Clapton) Ltd over a three-week stay at Aldenham using Silexine paints and varnish.

Above **Perhaps because its bold and largely uncluttered colour scheme related to the name of the product, the** *Yellow Pages* **Routemaster became particularly well known and was a strong advertisement for the power of an all-over message sensibly applied. Liverpool Street is the setting in this April 1971 view of RM 971.**
Colin Brown

Silexine RM 1737 created a sensation when it first appeared on route 11, but by March 1970 when photographed at Shepherds Bush the novelty had begun to wear off. The operating garage is Riverside although the code is not carried, nor is a fleet number.
Colin Brown

The complex tartan pattern on Scottish & Newcastle Breweries' RML 2701 was a superb example of the coachpainter's art. Photographed soon after receiving its livery in May 1972, it later ran in the same form from Highgate and Hackney garages.
Colin Brown

Ladbroke's cartoon-style scheme for RML 2547 provided for passengers to complete the headless jockeys and dog handlers. The vehicle ran from Hackney garage for most of its eight-month advertising stint.
Capital Transport

RM 1737 ran in its Silexine colours until 30th September 1970. Everything returned to normal until April 1971 when RM 971 burst upon the scene painted, appropriately enough, in a vivid canary yellow advertising Yellow Pages. As before, the vehicle attracted tremendous interest and comment which grew stronger in July 1971 when RML 2702 presented itself in orange, black and white under the banner of Homepride Flour. Soon advertisers were almost lining up to display their wares on London buses and the years 1972 and 1973 saw an outbreak of all-over adverts guaranteeing their appearance at one time or another in almost every corner of the system. In some instances it was a specific inclusion in the contract that they should move between garages and routes at pre-defined dates to ensure widest possible visual coverage. The Routemasters concerned, in order of appearance as advertising buses, were as follows:—

	Advertiser	Operated From	To	Basic colours
RM 1737	Silexine Paints	August 1969	September 1970	Pale green
RM 971	Yellow Pages	April 1971	April 1973	Yellow
RML 2702	Homepride Flour	July 1971	October 1971	Orange, black and white
RM 249	British Leyland Unipart	May 1972	May 1973	White and black
RML 2701	Youngers'	May 1972	August 1973	Tartan
RML 2547	Ladbroke's	May 1972	January 1973	Green and blue
RM 1270	Sharp Electronics	May 1972	January 1974	Violet
RM 2140	Bertorelli Ice Cream	June 1972	July 1973	Mauve and red
RM 686	Vernon's Pools	August 1972	September 1974	Blue
RML 2302	Evening News	October 1972	November 1973	Yellow, red and black
RM 762	Esso Blue	November 1972	December 1973	Light blue
RM 783	Esso Uniflow	December 1972	January 1974	Grey, green, blue and cream
RM 786	Ladbroke's	February 1973	October 1973	Yellow, green, blue and brown
RM 1740	Danone Yoghurt	March 1973	September 1973	Green and sky blue
RML 2280(1)	Hanimex	March 1973	October 1973	Blue and yellow
RM 294	Celebrity Travel Agents	April 1973	September 1973	Silver
RM 682	Pye	May 1973	December 1973	White and orange
RM 906	Barker & Dobson	May 1973	February 1975	Black and white
RM 952	Meccano (Dinky Toys)	August 1973	February 1975	Blue
RM 1285	Peter Dominic	September 1973	November 1974	Blue, white and orange
RM 1015	Austrian Wines	October 1973	April 1974	Red and orange
RML 2560	Ladbroke's	October 1973	October 1974	Green, blue and yellow
RML 2280(2)	Myson's	November 1973	December 1975	Blue and white
RM 1196	Myson's	December 1973	January 1976	Light blue and white
RM 995	Bank of Cyprus	December 1973	May 1974	White, blue and yellow
RM 1359	Boulogne Chamber of Trade	April 1974	September 1974	Yellow
RM 1255	Rand Office Bureau	June 1974	September 1975	Primrose and purple

With a total of 24 projects employing 23 buses (RML 2280 was used twice), the two years from May 1972 were particularly busy and colourful ones – at least from an advertising agent's point of view. Like so many good schemes, it was probably over-sold; certainly public interest began to wane as garishly-painted Routemasters became increasingly commonplace. In August 1975, when only the two Myson vehicles (RM 1196, RML 2280) and the Rand one (RM 1255), remained, it was officially announced that no more would follow. There had always been a strong lobby against what was seen as a debasement of the London bus livery, and now it prevailed. Except that just over a month later, on 2nd October 1975, RM 1676 appeared in cream and green promoting English Apples and Pears. Presumably the policy reviewers hadn't been told about this contract! It expired in April 1976, four months after a second

announcement had been made to the effect that there would definitely be no more this time.

RM 1237 was the next all-over advert and it did not appear until a decent lapse of time had occurred, in November 1979. A rather dull, mainly red scheme proclaimed Wisdom toothbrushes up to January 1981. In January 1983 RM 757 was painted blue and white as a trial advert for British Airways but did not run. Next came RMLs 2412 and 2444 which were contracted to Lyle & Scott from December 1983, the first in grey advertising jockey shorts and the other in white for Y-fronts, subject material which provoked criticism in some quarters. Both reappeared in standard red livery in March 1985. Last in the long line was RML 2492 which ran in a pleasant light green and yellow on behalf of Underwood's film processing from June 1984 to August 1985.

Top Left A nearside view of the one that started it all, RM 1737, at Aldwych in May 1970. Alan B. Cross

Left Detail views of the Yellow Pages and Myson RMs. Note the neat positioning of the legal lettering on the zebra crossing of the Myson vehicle. G.F. Walker

Above Spurning the use of a multi-coloured scheme, Barker & Dobson used the black and white 'livery' of its Everton Mints on RM 906. John Gascoine

Facing Page **RM 1740's link with yoghurt was one of the shortest of all overall advertising contracts, lasting only seven months, all of which were spent operating from Tottenham. In contrast, Bertorelli's RM 2140 (with fleet number on the driver's door) ran from four locations of which Battersea was the first.** Colin Brown

This Page **The last of Ladbroke's three advertising forays was on Upton Park's RML 2560. A much later mobile advertisement was Lyle & Scott's Y-Front RML 2444, also at Upton Park. The last in the line (so far) was RML 2492 advertising a printing and developing service and operating from Ash Grove garage on routes 6 and 11.** Colin Brown/G.A. Rixon/Ramon Hefford

CHAPTER FIVE
THE XRM – A DREAM UNFULFILLED

The low key launch of FRM 1 in December 1966 was linked to a mood of pessimism through the certain knowledge that it was already outmoded and would not go into production. Strong forces were at work advocating that London Transport should abandon all aspirations for designing its own vehicles and concentrate on buying manufacturers' standard models as all other operators were now doing. The advent of the New Bus Grant scheme, with its ready and hefty contributions towards purchasing new vehicles of approved design, further militated against the design and manufacture of specialist and perhaps unorthodox buses. London Transport plunged headlong down the 'off-the-peg' road amidst warnings of dire consequences which would surely flow from assuming that buses suitable for city use elsewhere in the country would also prove adequate in London. The subsequent disasters with Merlins, Swifts and Fleetlines were seized upon as confirmation that off-the-peg buses were less than satisfactory and there is no doubt that the very robustness of the Routemaster tended to highlight deficiencies in the newer models whose increase of some 35% in day to day maintenance costs was a heavy price to pay. However, as always, there were two sides to the story. The mythology that no one else could design buses for London tended to erect a bias against the new vehicles even before they entered service, and the prophesy

that they would fail then tended to become self fulfilling. London Transport was itself far from blameless with its abject unwillingness to modify its engineering practices to accommodate the new designs and its dogmatic adherence for a further decade and a half to the concept of unit and component maintenance at a central works rather than undertaking individual bus maintenance at garages to which the new types – which had not been designed to accommodate sophisticated flow-line overhaul methods – were far better suited.

Despite the forces seemingly ranged against the concept, the dream by the development engineers of designing a new generation of London bus remained constant. Even at the FRM launch the view was expressed that information gained would probably be used for a 36ft double decker of still more sophisticated design which was likely to materialise before 1969. The Transport (London) Act 1969 removed the Executive's powers to design and build its own vehicles but the satisfactory progress through Parliament in 1975 of the London Transport (Additional Powers) Bill restored this authority and paved the way for the XRM. Although London Transport was by now heavily involved with Leyland in the B15 project, later to reach its conclusion with the Leyland Titan, its own parallel independent design and research work forged ahead.

A much publicised drawing of the XRM in its originally intended form showing the position of the twin sets of low profile wheels in relation to the doorways. The 'tumble home' to the body sides was a design feature then all but abandoned elsewhere which, with the squareness of the upper bodywork, would have resulted in an ungainly, old-fashioned looking vehicle. An alternative drawing, identical in all other respects, shows the fitment of opening windows with which, after the failure of the forced air ventilation on FRM 1, the vehicle would almost certainly have been fitted.

The official coding of the project as XRM was an obvious choice; X for Experimental and RM because all that was best from the Routemaster, such as its superbly strong and light structure, would form the basis around which the new design would evolve. Once in production, the first task for the XRM would be to replace the RM fleet and herein lay one of the main imponderables facing the design team; would the RMs be replaced by driver-only vehicles or would crew operation still be in force? London Transport's desire for moving to complete omo looked difficult to achieve in central London, although high hopes rested with a new ticketing experiment in Havering aimed at reducing boarding speeds on driver-only buses to 2 seconds per passenger or less which would turn the key to the total elimination of conductors. Meanwhile design work on the XRM would have to proceed on the basis that it may be needed for either mode.

Above **In the cramped conditions of the London Transport experimental shop at Chiswick a mock-up of a four-axle double decker suitable for automatic fare collection takes shape in 1968. These were London Transport's first thoughts on what became the XRM project. The lower saloon layout was substantially complete, enabling the low floor, wide doorways and seating and stanchion positions to be clearly demonstrated. The inset rear window indicates that, at this early design stage, a rear-mounted engine was considered likely. A front view of the mock-up, at this time strangely code-named the 'International', appears in Volume One.** London Transport

Below Right **In 1975, the first full year of the XRM project, Leyland's B15 was unveiled, in which London Transport expertise was also heavily involved and where a conflict of interest may have been perceived to arise. Peckham's RM 1707 passes the first B15 prototype at its press launch. NHG732P saw trial service in London on the 24 during the period of crew DM operation of the route.** Capital Transport

Work which had been going on in a small way developed into a full scale project in 1975, funded with GLC approval under capital expenditure authorisation covering the next three years, with further allocations granted annually up to 1980 to carry the project through to 1983. Towards the end of 1975 the full scale of the XRM project became public knowledge. The extent of the new design impressed many within the industry as to the depth of innovation proposed but at the same time aroused scepticism over the practicality and cost of achieving much that was planned. The major design features of the XRM were as follows:

1. A side mounted engine (reminiscent of the old AEC Q type) to give total flexibility of door layouts.
2. A multi-axle, small-wheel layout to give the lowest possible floor.
3. A flexible, hydraulic drive system to replace the conventional transmission shaft, which was essential if the floor was to be substantially lowered.

Planned dimensions were 31ft 2ins (9.5m) length with a wheelbase of 16ft 2ins and overall height of 14ft 4ins giving a very generous headroom of 6ft 3ins in the lower deck and 6ft 2ins upstairs. A variety of different layouts had to be provided for, including single front entrance and exit, front entrance and centre exit, front entrance and rear exit, and front entrance with dual exits at centre and rear. The versions incorporating a rear doorway necessitated a twin staircase layout with consequent loss of passenger carrying capacity, but a single staircase variant could be expected to accommodate about 67 seated passengers (ie fewer than a DMS) with 8 standing passengers in crew operated mode or 19 without a conductor. Most of the seating in the lower saloon, with the exception of two transverse rows at the back, would be inward facing because of the long wheel arches front and rear.

The multi-wheel, low floor concept was not a new one but it was virtually untried. Two fairly recent attempts by other parties had both got some way before floundering but the lessons from them were studied by the XRM project team. In 1968 Leyland had undertaken a design study for a four axle Commutabus and had got to the stage of building an impressive and futuristic mock-up before abandoning the project because of problems over the transmission line and the small tyres. Even more ambitious had been a much-publicised eight wheeler built in about 1971 by Moulton Developments of Bradford-upon-Avon at the request of Sir George Harriman, then Chairman of the British Motor Corporation. Designed by Dr Alex Moulton of rubber suspension and bicycle fame, the Perkins V8-powered coach incorporated independently-sprung wheels interconnected with fluid as in the Hydrolastic suspension of BMC cars such as the famous Mini, with everything mounted on a lattice

Two interesting but totally unsuccessful projects, the concept of which had paved the way for the XRM in the use of ultra low floors and eight small wheels, were the Moulton super-coach and Leyland's single-deck Commutabus, the latter getting as far as the mock-up stage only. The Moulton vehicle is seen on test with that company. Moulton Developments Ltd.

structure of steel tubes of immense stiffness combined with low weight. As with Leyland's Commutabus, the will-power and financial resources to take a promising design through to its conclusion were lacking, but Dr Moulton was involved in design exercises on the XRM.

In order to study the problems associated with small wheels a Bedford VAL coach was purchased in June 1975 which became a familiar sight at Chiswick for a number of years languishing outside the experimental shop. RUW990E was a Plaxton bodied VAL14 variant of the famous twin front axle design which had made its debut back in 1962 at the same Earl's Court show as RMF 1254; it came third hand to London Transport having originated in 1967 with Homerton Coaches Ltd before passing in 1972 to

Sampson's Coaches & Travel Ltd of Cheshunt. Possession of the VAL brought the opportunity to study the steering features of multi-axled vehicles and to assess braking efficiency, the latter having not been one of the strong points of the VAL when first introduced. An immediate problem with a small-wheeled eight wheeler was that no suitable tyres were available anywhere on the market. Research work showed that, even in the event of such tyres being manufactured (and this possibility was not discounted as at least one manufacturer had shown interest), tyre wear was likely to be excessively high. The problem of tyre scrub which had been ever present on the trolleybus fleet would reassert itself with the double rear axles, and though it could be overcome by having a steered rear axle this was an unacceptable complication.

The Chiswick experimental shop makes an apt backdrop to Bedford VAL coach RUW990E standing head-on to DM 1787, the two main guinea pigs in the XRM project in its earlier days. Further along is MS 2, another long term resident at the experimental department. Ken Blacker

In order to obtain a really low floor – the target being a single 10ins step from the ground with a further 7ins step into the interior – the conventional transmission shaft had to be eliminated, and London Transport's thoughts turned to the concept of Hydrostatic drive. Developed by Donald Firth in conjunction with the National Engineering Laboratory at East Kilbride, this involved the engine powering a hydraulic pump connected by pipes carrying oil at high pressure to sophisticated rotors in each axle. Drive shaft, gearbox and differential were all eliminated. Though there may be loss in transmission efficiency compared with the conventional design, it was thought that this would be more than offset by energy storage through the rotors during braking, compressing the oil to subsequently provide motive power for acceleration. The engine could, in effect, be situated anywhere on a vehicle and this suited London Transport's objective of moving it away from the rear to the offside. The NEL, a government sponsored body, had been working on hydraulic transmissions since 1962 and London Transport had been in close touch with it since 1972. It was arranged that a vehicle would be sent to East Kilbride for conversion to hydrostatic drive and the bus chosen for the purpose was Fleetline DM 1787. The conversion proved more difficult and time consuming than anticipated as the laboratory's engineers struggled to obtain a satisfactory road performance from the vehicle. To obtain adequate acceleration four ball motors had to be mounted in each rear hub, two of which were cut off from the oil flow at about 20mph to increase the flow to the remaining two. For convenience the existing back axle was retained although the differential was removed saving some 300lb in weight; however this was more than restored by the heavier hydraulic pump fitted in place of the

gearbox. When the vehicle finally returned to London, much delayed, in November 1977 it was unable to obtain a satisfactory operating speed, and to make matters worse fuel economy had fallen dramatically from 6½mpg down to mere 4½mpg. At this stage the development was still incomplete and lacked the energy storage system, but even if it had been included it would by no means have made up the shortfall in fuel economy. The result must have been a huge disappointment to Sinclair Cunningham, the NEL project leader, who a year earlier had asserted that 'the hydraulic transmission shouldn't be any less efficient than the old transmission.' To the XRM project team it signalled an unsatisfactory conclusion for much of their work to date. The XRM had a production deadline of 1985 and too much work still remained on hydraulic drive systems for this to be met. This feature had therefore to be abandoned and with it the whole low floor, multi-wheel concept. In June 1978 it was formally admitted that the four axle design had been abandoned and that work was now being concentrated on a two-axle variant with normal sized wheels.

Almost overnight the XRM had changed from being a highly revolutionary concept to only a mildly different variant on an established theme. London Transport was now taking delivery of its first production batches of Titans and Metrobuses, which were regarded as second generation rear-engined double deckers and should be free of many of the problems associated with the Fleetlines. And indeed after modifications, particularly on the Titans, both models established themselves as reliable and reasonably economical to run. Doubts once again arose; was the XRM necessary when the Titan and Metrobus could be used as the basis for future developments? Furthermore London Transport was now publicly commit-

ted to dual sourcing for new vehicles after its unfortunate troubles with the Leyland group in the past, and such would be the economics of manufacturing XRMs that dual sourcing would be out of the question. However the new Ts and Ms still had to prove themselves and design work pressed on with the intention of producing two prototypes for testing in 1982.

One feature remaining unique to the XRM was its centrally situated, vertically mounted offside engine. Chosen for flexibility of body layout (it could sit under a staircase), it would be better sited for cooling than a rear-mounted engine and would be within earshot of the driver and therefore less open to unwitting abuse. It proved no easy matter to locate a suitable engine but eventually a Mercedes V-shaped unit was chosen as being the best, though doubts were harboured as to whether the GLC would countenance buying 'foreign'. The next best alternative appeared to be the proprietory in-line Leyland L11 but this would require modification to place its auxiliaries on the opposite side. However another problem now arose. The possibility of a shorter XRM of 28ft 8½ins (8.75m) length with a wheelbase of 14ft 6ins to give greater manoeuvrability on in-town routes was mooted and had to be investigated, but for this the L11 engine was too long and therefore out of the question. Fuel economy being one of the objectives of the XRM, experiments into the use of liquid petroleum gas (LPG) were carried out on single deck RM 1368 but were shown to produce no advantage either in economy or performance. RM 1368 was also employed in trials with modified power assistance for the steering system using an accumulator to provide oil under pressure only when required in preference to having fluid circulating wastefully around the system whenever the engine was running. This worked fairly well.

An artist's impression of one version of the 31ft 2ins XRM after the multi-axle project had been abandoned. Other versions were generally similar but covered a variety of door positions. Noteworthy is the use of a flat windscreen set at an angle as an alternative to the usual curved glass; a later version showed a sloping windscreen only on the offside, the nearside one being upright to accommodate the folding doors.

It was to the question of suspension that much time and effort was devoted. With a Routemaster-type hydraulic braking system and the use of hydraulics for all auxiliaries, the XRM would have no need for an air system and therefore air suspension was out of the question. However the ability to raise and lower the platform to speed boarding times was held to be necessary, and this meant again using hydraulics. For some time Automotive Products Ltd at Leamington Spa had been developing its Lockheed Active Ride Control (ARC) system using oil as an alternative to air as a suspension medium, for which the Company made the claim that almost complete elimination of roll and pitch could be achieved through the ability to respond very rapidly to changing load inputs. With London Transport's encouragement, RM 1 had been acquired and was fitted out as the ARC guinea pig at the

RM 116 in operation from Mortlake garage on route 9 in 1981 with the hydrostatic suspension in place and a diagram of the system as applied to this vehicle and, earlier, RM 1.
Capital Transport

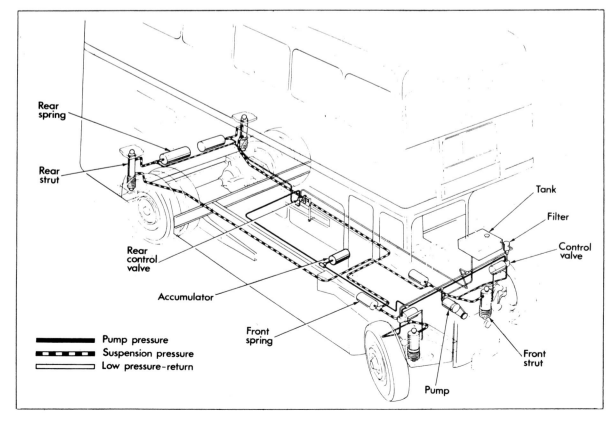

Right **The changes to RM 116 were all hidden by the bodywork and there was no external clue that the vehicle was in any way different from others in the fleet except for the rear corner flap behind which the fluid reservoir was located.**
LT Museum 80/450

Chiswick experimental shop under the supervision of Automotive Products' engineers. Initial road testing took place in March 1975 in which RM 2 and RT 365 also participated to give comparison with proven coil and leaf suspensions.

Active Ride Control employed hydraulic struts in replacement for the conventional Routemaster coil springs and shock absorbers, a pump and reservoir being used to vary the amounts in the struts. The pump was belt driven straight from the transmission line with the reservoir located under the bonnet. The idea of the system was to vary the amount of fluid in the struts according to road conditions and passenger loadings, giving the suspension a self levelling capacity. An accumulator was also fitted, under the floor, to damp out peak pressures from the pump or sudden pressures such as fast cornering or heavy braking. The design team was sufficiently impressed to give the system a service trial and a standard Routemaster, RM 116, was taken to Chiswick in February 1978 for this purpose. The ARC equipment was removed from RM 1 and fitted to RM 116 although with the difference that

a new hydraulic pump with straight-through drive from the alternator drive line was fitted in place of the original belt-driven arrangement. Service operation commenced from Mortlake garage on route 9 in June 1980 but breakdowns were frequent because the noisy hydraulic suspension pump drive, working in tandem with the alternator, caused the batteries to drain. This was subsequently modified and RM 116 continued at Mortlake, and later Stamford Brook, on and off on route 9 until its final demise in July 1987. It was never, however, provided with the facility, once planned, for lowering the platform. Considerable wear within the hydraulic struts, and fairly constant oil seepage from them, indicated that more design work was probably desirable, whilst the ride was never all that might be hoped for. Though, from a driving point of view, the vehicle cornered well and much of the characteristic Routemaster pitching was eliminated, passengers (especially on the top deck) were treated to a strange motion on sharp cornering as RM 116 first leaned and was then abruptly pushed upright when the Active Ride Control took effect.

Development work on the XRM was still in progress in 1980 but its future now hung in the balance. Failure of the Havering ticketing scheme indicated that 100% omo would probably not now be possible for many years to come unless some further unexpected breakthrough occurred to speed fare collection. The Titan and Metrobus were now reliable enough to see one-man operation through for a number of years and were both capable of development into a third generation of omo double deckers if necessary. As for the replacement of Routemasters, it was now the view that these could go on indefinitely subject to minor re-engineering, and the fact was that their design could not be bettered for the role which they were to continue to play. Doored buses had proved a failure on crew-operated routes, increasing journey times unacceptably and deterring trade, and although an open rear platform XRM had been considered the feeling current at the time was that under the present-day Construction & Use and Health & Safety at Work regulations, manufacture of new buses of this type would probably not be permitted. There was also the price factor to consider. The originally planned run of 2500 XRMs from 1985 would have required (at 1980 prices) a tooling-up cost of £9m plus a purchase price of £144m, a total of £153m. This contrasted with an outlay of £131.5m to buy 2500 Titans and/or Metrobuses or about £13.5m to rehabilitate the 2700 Routemasters for continued operation. It was accepted that, with its higher cost and marginal benefits, the XRM would be unlikely to find sales outside London to defray the development charges.

And so, in September 1980, the XRM was put to rest. The door to future development was not totally closed; in official words 'The idea of a bus designed by London Transport will be kept alive but much depends upon future technology in fuels and materials'. RUW990E was sold as redundant in June 1981 but, as already mentioned, RM 116's erratic progress was monitored for some years afterwards. Certain aspects from the programme have seen the light of day in other forms. The concept of a front entrance, rear exit (with doors) two-staircase crew-operated bus, which was one of the XRM alternatives, came to fruition when the Alternative Vehicle Experiment (AVE) programme of the mid-eighties found Ailsa Volvo V 3 at work in this form on routes 77A and 88 from Stockwell garage in 1985/6. Another Alexander-bodied Volvo, Citybus-type C 1, demonstrated an energy storage system harnessing braking power on route 102 from Palmers Green in 1986/7. Since neither was a success, it was perhaps fortunate that no firm commitment to pursue either course was made when the XRM project was still alive. However, despite the closure of the Chiswick experimental shop, the quest for a Routemaster replacement was not quite dead. Work on a new design was commissioned from three manufacturers, Northern Counties, Alexander and Dennis, and drawings were produced in 1990. Although not strictly within the scope of this volume, and ultimately abandoned anyway, it is included for completeness and outlined in Appendix 1.

CHAPTER SIX

NORTHERN GENERAL DEVELOPMENTS

Above **The corporate livery era has arrived and standards on Northern appear to be on the decline. Unlike the Northern Atlantean and United Lodekka now in poppy red, Routemaster 2119 still retains old colours apart from standard grey wheels, but its wheel trims and polished radiator grille have all vanished. The location is West Hartlepool bus station.** Ken Blacker

At the start of the 1970s Northern General's Routemaster fleet was rostered to cover many of the company's front line duties and these fine vehicles were, without doubt, the Rolls-Royces amongst buses in the North East. Sad to say, a general run-down in appearance became the trend in the new decade, hastened by the removal of the front and rear wheel trims and even, in many cases, of the polished surround from the front grille. The arrival of the National Bus Company corporate livery in 1972 only worsened matters. The drabness of the obligatory poppy red and white did much to depress the appearance of and pride in most fleets, particularly in its earlier applications when it quickly adopted an appearance like orange peel or faded metal primer. The logic behind it was never clear as the NBC's stated intention of retaining and fostering locally orientated bus companies ran totally contrary to the adoption of an all embracing corporate image. Routemasters began to appear in poppy red from late 1972 onwards although a

slowing down of the repainting programme meant that several lasted into 1975 before succumbing. In that year, two Routemasters were amongst many vehicles in the Northern fleet to receive Tyne & Wear PTE yellow applied in NBC style with a single white band and they ran in this unusual garb for a few years.

One of the more unusual experiments of recent times took place in 1972 on the lowest numbered Routemaster, 2085, which had suffered severe front end damage whilst working a Sunderland local service in April of that year. In common with operators throughout the country, Northern was experiencing serious reliability problems with its rear engined double deckers and, under its Chief Engineer, Mr D.A. Cox, it set out to assess the feasibility of converting existing front engined double deckers to a layout suitable for one-man-operation. The first conversion used a 1958 Leyland PD3, because the company did not want to risk 'one of its precious Routemasters', which was extensively rebuilt from rear to

front entrance and, at the same time, converted to a normal control configuration by pushing the driver's position and the whole of the front bodywork backwards by about 2ft 6ins. Separate entry and exit doors were provided adjacent to each other at the front of the bus with the entry position immediately facing the driver who now sat inside the saloon in the manner customary on any normal control vehicle. Despite early official publicity to the contrary, the Tynesider as this bus was named, was not a success, the extensive and costly rebuilding having completely altered the weight distribution and handling characteristics which reflected particularly badly in the steering. The second experiment, in which the Routemaster participated, was less ambitious but just as interesting.

The same principle of moving the driving position backwards was applied although, this time, the distance involved was only about 1ft 6ins. In this case the main upper saloon body structure was left intact, as was the lower deck from the entrance door backwards. The

Right **The 'Wearsider', alias 2085, was rebuilt with very little alteration to the nearside of the Routemaster body upon which all windows and the passenger doorway remained in place. The 'Wearsider' plate at the top of the front grille is of interest as is the fact that, through moving the windscreen back, the periscope glass for the indicator box is now outside the cab.** Ken Blacker

Centre **An offside view of 2085 shows the alterations made to accommodate the repositioned cab and staircase. Photographed crossing Monkwearmouth Bridge into Sunderland during its first weeks in service, most of its time in this rebuilt form it was used on duties other than stage work.** G. Dudley

Bottom **At the rear of Northern's Washington depot, 2085 stands alongside the Tynesider later in life. The 1958 Titan PD3 conversion was re-registered when rebuilt with Routemaster style bonnet in 1972.** Steve Warburton

conventional bulkhead had to be removed to accommodate the new driving position which now faced the doorway, and the staircase was set back by the same amount as the driver's seat. This resulted in the loss of two seats downstairs and the removal upstairs of one pair of seats from behind the staircase to in front of it. Two single seats facing the stair-well were converted to double to compensate for the loss of the two seats downstairs with the result that the total capacity remained at 72. The rebuilding involved the loss of the front air intake and heater exchange unit following from which the standard Route-master fresh air heating system was replaced by a Webasto paraffin unit located under the staircase. The vehicle was re-upholstered in the company's 'Northern Rose' moquette and repainted externally in an attractive red and white livery already adopted for the Tyne-sider; in this instance the vehicle was chris-tened Wearsider with the name written prominently in script on the between-decks panels. However, despite the handsome livery, 2085 looked a mess. The normal good looks of the Routemaster were destroyed by the mass-ive overhang of the upper deck forward of the repositioned windscreen, and the bonnet snout resulting from its semi-normal control layout combined with the retention of the half width cab to produce a design hotch-potch.

The cost of rebuilding 2085 at Bensham worked out at about £1250. In August 1972, it entered service back at Sunderland but, before long, was transferred to Jarrow depot for what transpired to be its only spell of one-man-operation, which was short lived. November 1972 found it at Washington where it remained for the rest of its working days, though not an active participant in the main-stream of that depot's Routemaster opera-tions. Subsequent repainting in corporate poppy red livery enhanced the appearance of 2085, now renumbered 3069, even less and also eliminated the Wearsider name. The end of the road came in July 1978. It is interesting to recall a statement made by the company, when the Wearsider was created, to the effect that, if reaction from staff and public was favourable, as seemed likely, the remaining fifty Routemasters would be similarly dealt with. Fate decreed otherwise.

No.3106 (formerly 2122) in full NBC corporate livery at Newcastle Worswick Street bus station. The route numbering, six hundred above the old Northern series, is a result of NBC corporate policy.
John Fozard

January 1975 saw the renumbering of the whole Northern fleet, when the Routemasters became Nos. 3069-3118, keeping the same sequence as before, with the former RMF 1254 now numbered 3129. Later in the same year, 3129 went through its last overhaul at which it gained a number of Northern's own style window pans, incorporating sliding ventilators, as replacements for original winding units. It had already lost the opening windscreen in favour of a fixed unit and, alas, lost its commemorative plaque. In the following year, it also lost its unique London type rear axle following mechanical failure, receiving in replacement a standard unit from one of Northern's own fleet which had already been withdrawn.

The first Routemaster withdrawal was a one-off resulting from a serious accident involving 3093 in October 1976. However, this predated, by only a few months, the commencement of the formal withdrawal programme which started in May 1977 and, by the end of 1978, left only 22 buses in service. Withdrawals slowed in 1979 with the result that 18 were still operational at the start of the nineteen-eighties but not for much longer. One of the last survivors was the ex-London bus which lasted through to October 1980. The last one of all, 3105, ended the Routemaster era on Tyneside as it had begun, with a press call for the local newspapers, on Tuesday 16th December. Later the same day, it worked from Leam Lane Estate into Newcastle, arriving there shortly after 6pm to bring an interesting era to an end.

In retrospect, the Routemasters on Tyneside marked a fascinating diversion from run of the mill bus development of the time, but little more. After purchasing the fifty, Northern turned once more to its favoured Atlanteans. Without doubt, they served the community well, and wore well too; at their first Certificate of Fitness recertification in 1970-72 the company stated that the absence of corrosion within the body frame meant that the cost of overhaul per bus was £1000 less than for a contemporary Atlantean but, at the end of the day, Northern's Routemasters lasted in service no longer than those same Atlanteans. Indeed, a few Atlanteans purchased before the Routemasters even outlasted them. The potential for longevity was never achieved; with the unstoppable onslaught of one-man operation they were bound to fall prey to early obsolescence. Their final days on Tyneside coincided with a decline in the general appearance of the once smart Northern fleet and the Routemasters often looked unkempt and uncared for. Internally, cheap looking plastic replacement seat coverings were a far cry from the style of earlier days.

After withdrawal, the majority inevitably found their way to the scrapyards. Two, 3074 and 3101, were actually sold to London Transport in January 1978 for the spare parts that they would yield and were broken up on their behalf by Wombwell Diesels. Twelve more were purchased by London Transport in 1979/80 with a view to possible further use but, as outlined in a later chapter, this came to nothing at the time although a lone example, converted to open top and purchased much later, was in active service with London Buses Ltd at the time of writing. In fact, comparatively few have seen further use, as PSVs or otherwise, since leaving Northern but, in chapter 15, we give a small insight into the subsequent story of those few. Fortunately, several were acquired by preservationists and even though some of these have since been scrapped following the collapse of well-intended schemes, a few will remain as a reminder of Northern and its Routemasters.

RMF 1254 early in 1975 shortly after being renumbered from 2145 to 3129 and, left, soon after its final overhaul. A one piece windscreen, reflective number plates and, in the second picture, a selection of sliding windows are apparent, and the bus has also received front wings standard to the rest of the company's Routemaster fleet.
John Gascoine
Paul Hulyer

CHAPTER SEVEN

AIRWAYS ROUTEMASTERS

As if to celebrate the new decade, a rapid repainting programme saw the whole 65-strong BEA Routemaster fleet carrying the new but unflattering orange and white livery by June 1970. Three years of relative stability ensued, but nothing in the airline world remained static for long and, on 1st September 1973, British European Airways came to the end of its existence when it was merged with BOAC to form British Airways, in which it became the European Division. Yet another new image was clearly called for, and little time was lost in applying the new corporate style on a rolling basis, starting with BEA 17 whose debut on 17th November saw a re-emergence of blue as the main colour. This time navy blue was used which, though quite a bit darker than the original blue, presented a much better appearance than the orange, though it now extended only to the top of the lower deck side panels, the remainder of the bodywork being white. Over the ensuing fifteen months, a further 51 Routemasters were repainted in the British Airways image, leaving thirteen in orange. These were BEAs 21, 26, 27, 29, 35, 38-40, 46-48, 52, 56. The reason for not repainting them was that they were no longer required. Three had already been withdrawn from service, commencing

with BEA 48 on 1st January 1975 and others were to follow, culminating in the sale of all thirteen to London Transport on 29th August 1975.

Not only were the airlines changing their image, but their whole method of operation was changing, both in the air and on the ground. Ever larger aircraft were being produced to cope with the massive surge in air travel, culminating in February 1969 with the Boeing 747 'Jumbo' jet capable of carrying well in excess of three hundred passengers. Heathrow Airport was in the process of being linked to central London by the Piccadilly Line tube, thereby casting a heavy shadow of doubt over the future of the dedicated coach services from London whose trade was diminishing. Quite some time beforehand, the concept of linking each coach with a specific flight had ceased and, instead, coaches operated on a regular timetabled basis with departures from Heathrow and Gloucester Road at regular intervals. Any attempt at taking trailers direct to the tarmac had also long been abandoned and they were unloaded by passengers on arrival at the Heathrow departure building. The Gloucester Road air terminal had ceased to serve as a customs check-in point and, under the new British Airways auspices, its very existence was under the microscope.

Their comparatively recent repaint into the new blue livery did not prevent a further fourteen Routemasters from being withdrawn at the end of the 1976 summer season, on 30th September, followed by their sale to London Transport on 9th November. 1977 was a year of stability but, between May and July 1978, a further nine withdrawals took place. There was no immediate purchaser for these and they were, initially, stored at Chiswick tram depot but tenure here was now almost at an end. The premises were required for conversion into Stamford Brook bus garage and, though it was now acknowledged that the airport coach operation would soon cease, a temporary home had meanwhile to be found for it. Stonebridge bus garage was the location selected but the nine withdrawn Routemasters could not be housed there and were removed for storage at Heathrow. Only 29 coaches, less than half the original fleet, moved to Stonebridge to commence operating from there on 6th August 1978. These were BEAs 1, 4, 6-10, 12, 19, 24, 25, 28, 31, 32, 34, 36, 37, 41, 43, 44, 51, 55, 57-59, 62-65.

Above **BEA 24 stands at London Airport in August 1973 carrying tangerine livery, and another four of the fleet are visible behind. In contrast is the new Locomotors trailer which has been delivered in the new British Airways blue.** J. Tilley

A 1976 view of BEA 25 arriving at Heathrow shows the final British Airways corporate livery which, despite the very large area of white, suited the lines of the Routemaster body well. Capital Transport

Towards the end of operation, the condition of the Routemaster coaches dropped dramatically. BEA 6 is disgracefully dirty and unkempt for what had always been regarded as a premium service, the cleanest part being the vinyl in-house advertisement on the side. Ken Blacker

On 19th February 1979 British Airways finally gave notice that operations would cease after Saturday 31st March 1979. This was much less than the three months' notice stipulated under the contract between the two parties but London Transport did not demur as they were glad to get back the staff who would be released. A planned recertification programme for the remaining Routemasters scheduled for the end of the year was immediately cancelled. BEA 25 had already failed to stay the course and was unlicensed at Stonebridge; the remaining 28 were delicensed when the operation ceased. The airport coach service which had once been so vital had become, in the airline's view, an expensive anachronism costing about £800,000 a year to provide, and its demise passed without protest. For the staff who manned the service, all of whom were seconded from the main London Transport fleet, the passing of this family-like operation was a sad event and about fifty were present to bid their last farewell to a passengerless BEA 59 as it set off for the final trip from Gloucester Road to the airport. On 5th June 1979 London Transport purchased all the vehicles which it did not already own, including the non-runners. The story of the 65 coaches as London Transport's RMA class is resumed in chapter 11.

CHAPTER EIGHT

PRESERVING THE PROTOTYPES

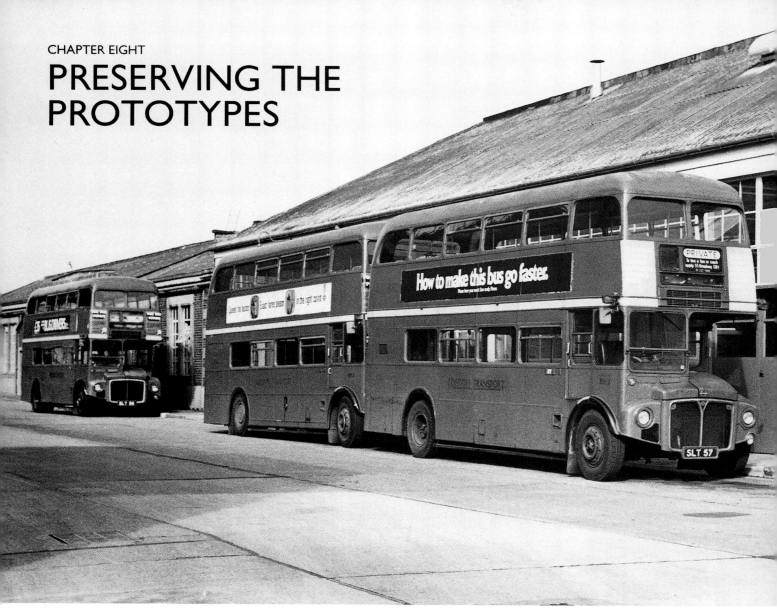

In the spring of 1973, the original Routemaster was sold to the Lockheed Hydraulic Brake Co Ltd of Leamington Spa for use as a mobile testbed for experimental purposes. Towards the end of March its fleet names, garage codes and legal ownership lettering were removed. Despite its merited place in bus history, there was no sentiment shown, nor does any thought appear to have been given to the possibility of preserving the vehicle. When it left London Transport on 19th April 1973, RM 1 was already almost nineteen years old and was widely regarded as being on its last lap of life prior to the inevitable scrapheap. However, it still appeared in the capital from time to time, becoming ever more scruffy, before ending up back at Chiswick Works, Lockheed no longer having any use for it. It remained resident for quite a while on the famous 'dip' which had formed part of the Chiswick internal test track in earlier times. Seventeen years on from its last repaint, RM 1 was understandably very down at heel when, on 14th July 1981, it was taken back into London Transport's stock, its historic value having at last been fully acknowledged.

Some general mechanical repair work and a fresh coat of paint applied at Aldenham transformed the pioneer Routemaster back into condition worthy of its new role as a show vehicle from May 1982 onwards. Because of cost considerations, the standard Routemaster front end of 1964 was retained, but there can always remain the hope that, one day, it may revert to its original form. In February 1986 the vehicle was handed over to the care of the London Transport Museum who, after long deliberation as to its future, finally officially absorbed it into the collection in August 1989. Space was not available for it to be displayed with the main collection at Covent Garden so it was put into storage, from which it is occasionally taken for display at special events.

From 15th February 1973, RM 2 was allocated to the experimental department at Chiswick and during 1974 received a platform door from an RT decimal currency trainer. It remained with London Transport and its subsequent exploits feature in later chapters. Since August 1986 RM 2's circumstances have paralleled those of RM 1 under the care of the London Transport Museum,

although RM 2 was placed on loan to the Oxford Bus Museum in August 1988.

RM 3 left the fleet in February 1974 when it was sold to the London Bus Preservation Group, at whose Cobham premises the vehicle is to be seen on open days. The story of RMC 4 after its transfer to London Country has been covered in chapter 2. The non-standard mechanical features of the prototypes mean that some ingenuity is needed at times when parts need to be replaced. It remains to be seen which of the four vehicles will be the first to have its bonnet assembly restored to original design.

Above **RMs 1 and 2 were withdrawn from training duties at Dalston and Highgate garages respectively in 1972. Both vehicles, together with RM 8, are seen at Chiswick awaiting a decision on their fate. Like many standard RMs at the time, the front end of RM 2 has become a hotch potch with both bonnet top bullseye and a grille mounted one, and with a brake aperture grille on the offside but not on the nearside.** John Gascoine

Looking forlorn and abandoned, RM 1 stood for a long time on the famous 'dip', part of the one-time test track within Chiswick works. It was photographed in August 1979, after its return from hydraulic system experiments with Lockheed but before being taken back into London Transport stock for subsequent renovation. SMD 62, standing behind and already withdrawn for disposal, represents a totally different and much less happy era in London's transport saga. Alan B. Cross

RMs 1 and 2 side by side outside the Covent Garden museum after external renovation by London Transport but before being officially taken into Museum stock. It is the hope that one day at least one of them will be able to revert to the front end layout of their service days.
Capital Transport

The first Routemaster to go into preservation was RM 3, photographed at the London Bus Preservation Trust's Cobham bus museum along with examples of the RT and RLH classes.
Alan B. Cross

CHAPTER NINE
THE QUEEN'S SILVER JUBILEE

The Silver Jubilee of Queen Elizabeth II and Prince Philip in 1977 presented an opportunity for celebration and self-publicity too good to be missed by London Transport. An excellent formula was hit upon; a fleet of buses would be painted in special Silver Jubilee attire and sponsors would be found to fund the project, the success of which would contribute handsomely to London Transport's coffers. In late 1975 advertisers were sought who each would be willing to pay £10,000 for the exclusive advertising rights on one vehicle, inside and out. As a trial run Chiswick's experimental RM 2 was painted in May 1976 in a livery of silver relieved by a red central band and a red bullseye in a forward position between decks. Soon afterwards standard RM 442 was

given the definitive treatment similar to RM 2 but with the addition of the stylised Crown and St Paul's logo adopted for the celebrations and the wording 'The Queen's Silver Jubilee Celebrations 1977'. Further adorned with lions rampant on either side of the front indicators, various small insignia at the rear, and advertising panels reading 'Our seventy-seven silver strip', RM 442 was all ready to demonstrate before committed advertisers and their agents at the Tower Hotel, St Katharine's Dock, on Tuesday 6th July 1976. A target had been set for the sponsorship of 25 silver Routemasters, and it was achieved.

It was decided to draw the silver fleet from newer Routemasters passing through Aldenham overhaul, and in January 1977 the decision was made to give them new, temporary fleet numbers in the SRM series. This was an unusual move and one which has never been repeated. The first completed vehicle to emerge from Aldenham was RM 1850 (alias SRM 25) on 18th February and the remainder quickly followed. The complete batch of SRMs was:—

SRM			
1	RM	1898	Abbey National
2	RM	1848	Addis
3	RM	1650	Air Jamaica
4	RM	1889	Amey Roadstone
5	RM	1668	Bulmers
6	RM	1912	Townsend Thoresen
7	RM	1871	Exide Batteries
8	RM	1787	Daily Mirror
9	RM	1907	Farley's Rusks
10	RM	1914	Goddard's
11	RM	1910	Heinz
12	RM	1911	ICL
13	RM	1648	Smirnoff
14	RM	1896	International Paints
15	RM	1903	JVC
16	RM	1920	Kleenex
17	RM	1894	Kosset
18	RM	1906	NatWest Bank
19	RM	1904	Nescafé
20	RM	1899	Avia Watches
21	RM	1870	Selfridges
22	RM	1900	Singer
23	RM	1902	Tate & Lyle
24	RM	1922	Lambert & Butler
25	RM	1850	Woolworth

Neither RM 2 (which stayed silver until required for the Shillibeer scheme in 1978 but remained elusive) nor RM 442 (which was repainted red in August 1976) ran in service as Silver Jubilee buses. The rest entered service on 11th April after first appearing in the Easter Parade on the previous day, during which SRMs 13, 14, 19 actually took part in the parade whilst the others stood on display in Battersea Park. Some twenty-one defined routes were reserved for the SRMs (although there were odd occasions when they strayed),

and in order to spread the advertising as widely as possible there were wholesale re-allocations of buses in June/July and August/September. For its trade appearance RM 442 had been equipped with light brown carpeting on each deck (which it retained in service at Streatham after being repainted red), and the main batch of 25 had very attractive carpets incorporating London Transport's Jubilee bus symbol, the Crown and St Paul's Insignia of the London Celebrations Committee, and the Woolmark. They were supplied free of charge

by the wool industry who used the opportunity to evaluate the wearing properties of 100% wool carpets under such arduous conditions. With one exception, National Westminster Bank, all advertisers paid for special yellow Gibson rolls to be used carrying their advertisements on the back. The advertising contracts ceased on 5th November and from the following day the vehicles began returning to Aldenham to receive their old liveries and fleet numbers back. The last one, SRM 4 at New Cross, ran on 30th November 1977.

Above **Two young boys take a closer look at RM 1922, alias SRM 24, at the Norwood terminus of route 2. The Lambert & Butler adverts, mostly black in colour, were among the more eye-catching designs on the silver painted buses.** Capital Transport

Left **In 1977 Heinz were still using the '57 Varieties' logo for their food products and their bus carried this potentially misleading advertising (shades of the tourist route publicity of some ten years in the future). SRM 11 is seen at Ladbroke Grove while working from Willesden garage.** John Gascoine

Above **Good use of colour made the Nescafé adverts on SRM 19 another striking design. The running number and garage code plate holders were painted black on the SRMs for the benefit of those garages still using stencilled unpainted aluminium plates, but Streatham was among those using white-on-red plastic plates.**
Capital Transport

Left **The 100% wool carpet fitted to the 25 SRMs carried this design by the International Wool Secretariat incorporating a London Transport Silver Jubilee logo, the crown and St Paul's symbol of the London Celebrations Committee and the Woolmark. The colouring was designed to match that of the Routemaster seating moquette. This photograph shows a length of unused carpet (centre) compared with carpet as removed from a bus at the end of 1977 (left) and used carpet after cleaning (right). IWS**

CHAPTER TEN
SHILLIBEER, SHOPS AND CELEBRATIONS

No doubt much encouraged by the success of the Jubilee fleet, London Transport grasped upon another opportunity to repeat the exercise when the 150th Anniversary of George Shillibeer's original London omnibus arose in 1979. Once again RM 2 served as guinea pig and this time its transformation was superb. On 21st February 1978 it was displayed for the convenience of advertisers outside Wembley Conference Centre in olive green and cream livery with a red central band, the fleet name 'Omnibus' in gold picked out in red, and beneath it the words 'George Shillibeer 1829 London Transport 1979'. Attractive symbols based on those carried originally by Shillibeer were positioned between decks, and the whole result was a harmoniously successful adaptation from his original. 'Advertise on our Omnibus' was the message which RM 2 carried and though London Transport had hoped to attract 25 clients at £10,000 a time it was reasonably happy at getting twelve (even if North Thames Gas ended up taking five vehicles, presumably at a discount!). Indeed there was a thirteenth sponsor who, being Leyland Vehicles, required something newer than a Routemaster and sponsored DM 2646. Once again the Routemasters selected for operational purposes were all newer ones, the twelve being RMs 2130, 2142, 2153, 2155, 2158, 2160, 2184, 2186, 2191, 2193, 2204, 2208. No renumbering took place this time; 'proper' fleet numbers were retained and, though gold, were of plain modern style and not red-shaded Victoriana as on RM 2.

Repainting into Shillibeer livery took place on overhaul, the ex-London Country exhibition and cinema bus RCL 2221 being similarly painted at the same time for its own debut in its new role. Tiny variations in paint style took place, mainly in the front wing and bonnet area where the cream picking-out responded to the types of unit fitted. The great day came on Friday 2nd March when the Shillibeer livery was launched in fine style at the Guildhall by the Poet Laureate Sir John Betjeman who, perhaps inappropriately in view of the massed ranks of green, cream and red motor buses on show, stated how much better things had been before motor buses came on the scene to replace horses. Revenue-earning service commenced the next day on a similar pattern to the Silver Jubilee fleet, a three-phased operation with changeovers in June and August/September. Several purely suburban routes were included within the schedule to give maximum advertising coverage, but beyond this, more unofficial working of non-prescribed routes took place than had occurred with the Jubilee fleet. The contracts expired on 30th November (although at its own request one advertiser, Esso, had backed out some three weeks earlier to coincide with the ending of a specific campaign nationwide), but there was no immediate rush to repaint the vehicles red. Apart from RMs 2186 and 2204 all were still green into 1980 although now carrying normal advertising matter and no longer route-bound. Repainting into normal livery was completed in February.

Attempts to get sponsors for the Shillibeer fleet must have been to some extent dissipated by concurrent searching for advertisers for a new operational venture which London Transport had decided to undertake. 'Shoplinker' was precisely what the title aptly described, a circular service from Marble Arch linking the main West End shops. Inaugurated under heavy pressure from the GLC but against the wisdom of many who doubted whether there was sufficient demand for a service which, in any case, was bound to be disrupted by traffic jams, the Shoplinker was set to operate from 7th April 1979. An eye-catching red and yellow livery was decided upon and at the end of 1978 RM 59 was suitably adorned with red lower panels and roof, the remainder being yellow. 'Our routes pass your branches' cried out the advertising material on the sides with a smaller and more discreet 'Advertise on Shoplinker' at the rear. Repainting of Routemasters into red and yellow took place simultaneously with the Shillibeers, although the overhaul process was complicated by the requirement to fit loudspeakers for the playing of onboard music and advertisements. A press launch on 18th

Above **A big event in Battersea Park on Easter Sunday 15th March saw the Shillibeer Routemasters interspersed with London bus relics of various ages as they paraded in front of large crowds of onlookers. Leading the parade was the replica of Shillibeer's original horse bus with RM 2191 following behind.** Capital Transport

March in Oxford Street found RMs 2146 and 2188 temporarily decked out with dummy shop windows, canopies and blinds. However the advertising take-up was not as good as hoped and approximately half the fleet carried in-house advertising when they entered service from Stockwell garage on 7th April, dissipating scarce staff resources and using vehicles desperately needed elsewhere. In this case the prototype Shoplinker was actually used in service, the 16-strong fleet consisting of RMs 59, 2139, 2146, 2151, 2154, 2159, 2162, 2163, 2167, 2171, 2172, 2174, 2187-89 and 2207.

The service did not prosper and for most of the time buses ran around almost empty. It was decided to discontinue the flat fare of 30p in favour of graduated and 'day' tickets and to operate the circle both ways round. These changes were due to commence on 28th July but were not, in the event, implemented even though the 'Fare 30p' message was eliminated from the destination screens. London Transport had reached the conclusion that the service could not be made a success, even with a planned extension to Victoria, and it was decided to cut losses by discontinuing it at the end of the summer. On the last day, 28th September 1979, RM 2188 closed down the operation with special destination screens reading 'Good Buy' and a 'Last Shoplinker' plaque on the front grille. As if to firmly close the curtain on this financially disastrous episode, the Shoplinker RMs were all immediately withdrawn for repainting and did not operate again in their special livery. Most were already back in red by the end of October and early November found them all restored to normal service. A few relics of the Shoplinker era survived, however, one of which was the microphone installation by which they could henceforth be identified from inside. RM 59 retained its yellow fleet numbers for a while (the other Shoplinkers had all carried standard white numbers) and three vehicles continued to carry special advertising material until contract expiry in April 1980. These were RMs 2187, 2154, 2207, the last two of which had wrap-round bands which sat very well on the red livery.

Top The beautiful green, red and cream Shillibeer livery graced RM 2 throughout most of 1978 in an endeavour to drum up trade from advertisers, with a fairly encouraging result for an event with potentially less public appeal than the royal celebrations of the year before. The superb signwriting included even the red shaded gold fleet number in ornate style. Mike Harris

Centre A launch for the Shillibeer liveried buses was held at Guildhall on 2nd March. Poet Laureate Sir John Betjeman arrived by horse bus, a much more civilised way of travel in his view. Capital Transport

Left A not totally convincing queue of passengers — in fact all are actors — waits outside Selfridges one weekend in March 1979 during the production of a television advert for the new Shoplinker service. Mike Harris

RCL 2221 makes its first major public appearance in Shillibeer livery at the 1979 Easter Parade rally in Battersea Park. In its new role as an exhibition and cinema bus it has already sustained minor bodywork damage. Capital Transport

'This bus goes to Leeds' claims RM 2186 at Camden Town. The floral wreaths each side of the advert are based on those carried each side of the 'Omnibus' fleetname on George Shillibeer's original vehicles. Capital Transport

Though a purely suburban operation, route 83 had the advertising advantage of running through the heart of the area served by North Thames Gas, sponsors of five Shillibeer RMs including RM 2155 seen at Wembley. Colin Brown

Right Lack of sufficient interest from potential advertisers meant that some of the Shop Linker RMs had to carry posters publicising the service itself. In plain black lettering on a white background these read 'Buy almost anything in London on this bus'. RM 2189, however, was one of the luckier ones, its advertising sites being shared by The Scotch House and Burberrys, stores of both companies being situated in Regent Street where the bus is seen. Capital Transport

Centre Two Shop Linker RMs carried attractive wrap-round adverts which were retained when the buses were repainted back into standard livery. RM 2207 publicised the HMV record store; the other vehicle, RM 2154, carried a similar style of advertising for Austin Reed, the gentlemen's outfitters. Colin Brown

Bottom 'Parcel' bus RM 520 takes a day off from its normal activities to operate the well patronised special tour on 28th July 1981, the eve of the royal wedding. The silver ribbon painted around the lower saloon and across the roof, and tied by a bow above the filler cap, holds a large greeting card from the sponsor at the rear. G.A. Rixon

Innovation in design inevitably carries the risk of being deemed gimmicky, but of the various designs sponsored by London Transport the one for the wedding on 29th July 1981 of Prince Charles and Lady Diana Spencer has probably fallen more firmly into this category than any others. It first appeared in March 1981 on RM 490 whose basic all-over red was dressed like a parcel with silver gift wrapping along the band below the lower saloon windows and also up and across the centre of the vehicle, including the roof, with a large bow outlined in black just above the filler cap on the offside and also on the equivalent second bay position nearside. A 'greeting card', painted as though tucked into the ribbon at the rear of the offside provided space for a message from the sponsoring company who also advertised on the main side panels. A minor modification was made internally so that the sponsor's posters could be covered in laminate to prevent defacement. Eight vehicles were sponsored and appeared from Aldenham in parcel livery – as it inevitably became known – in time to enter service on 13th June. They seemed to lack the gloss and fine finish of earlier special-liveried buses and the silver appeared in some lights more as a drab grey. RM 490 did not operate in service as a 'parcel' and was repainted back into standard livery in July. The eight sponsored buses were RMs 219, 519, 520, 534, 559, 561, 595 and 607.

Each bus was allocated to a separate garage and stayed there except in the case of RM 219, which moved from Streatham to Camberwell because of vandalism, and RM 519 which moved from Middle Row to Westbourne Park upon the closing of one and opening of the other. Although allocated to defined routes they often strayed and some garages appeared to have little interest in ensuring that they adhered to specific duties. On the day before the wedding, 28th July, the eight operated a special tour along the processional route and in the evening served Hyde Park for the Royal Fireworks Display. Such was the passenger demand that two ordinary RMs were called upon to help out. Thereafter they returned to normal service operation until called to Aldenham in September and October for repainting into normal livery.

In 1983, London Transport – so soon to be dismembered though no-one knew it at the time – proudly celebrated fifty years of existence. Official celebrations saw the repainting of four vehicles into special liveries of which one was a Routemaster. RM 1983 was repainted in April in all-over gold with white band, black mudguards and Jubilee emblems in place of roundels; it carried the slogan on both sides 'We've been together now for 50 years'. The other three were Titan T 747 which was also painted gold and temporarily renumbered T 1983, and M 57 and T 66 which were both repainted into 'General' livery. None of these depended upon sponsoring and indeed did not carry any commercial advertising. The Routemaster entered service at Croydon on 30th April and was scheduled to visit all eight operating districts during the summer although mechanical problems dogged the earlier part of its programme. After a final move to Clapham on 5th September it remained there, the gold turning ever more drab, until it went for repainting on 2nd February 1984.

This was an era of high initiative in garages evidenced by the numerous and successful 'showbuses' managed by staff at their own time and expense. This initiative reached its fulfilment in the total repainting of four RMs into the handsome red and white livery with black lining and silver roof in which London's buses ran in 1933. Sidcup's RM 8 was first in March 1983 followed over the ensuing weeks by RM 17 at Willesden, RM 1933 at Chalk Farm and RM 2116 at Seven Kings. There were minor differences between all four in interpretation of the original livery, notably in the area of cab and canopy, but each looked resplendent demonstrating how well a fifty-year old livery suited the Routemaster's classic lines. Minor changes were made in the summer to RM 2116 – which was then named Forest Ranger – and to RM 17 whose black-painted cab was replaced by white which, though not authentic by 1933 standards, suited it better. An assortment of other RMs received minor unofficial treatment such as gold bands in their showbus capacity, as described in another chapter. All four entered

1984 in 1933 livery but a management change of heart decreed that they – along with all other Showbuses – must revert to standard livery. Thus RMs 8 and 1933 became standard red in February and RM 17 in April. Only RM 2116 survived, to be delicensed on 21st May and subsequently sold to enthusiasts who, happily, have retained its 1933 colours.

Above **Gold was an appropriate colour with which to celebrate fifty years of London Transport in 1983. An obvious choice to spearhead the celebration, RM 1983 tended to look fine, almost sparkling, in bright sunshine but could appear drab in overcast or wet weather. Luckily the former prevailed in June when RM 1983 was photographed at Feltham.** G.A. Rixon

The 1933 livery chosen by a few garages as their own local commemoration of fifty years looked very handsome when applied to Routemasters. Chalk Farm was fortunate enough to possess RM 1933 which was a natural candidate for repainting in the colours of fifty years earlier. It is seen in Pimlico in September. Mike Harris

Below Left **RM 8 also carried 'General' colours, but with minor differences in application compared with RM 1933. It is seen on its first day in service with the special livery.** G.A. Rixon

Above **A few buses were given gold relief bands and white Golden Jubilee logos. RM 1091 is seen at Hyde Park Corner working from Riverside garage.** G.A. Rixon

Highly successful open days were held at both Chiswick and Aldenham works during Golden Jubilee year, the former on 2nd July and the latter on 25th September. Routemasters were strongly represented at both occasions — this line-up of showbuses is seen at the Chiswick event. Colin Brown

CHAPTER ELEVEN

LONDON'S SECONDHAND ROUTEMASTERS

In 1975 the inevitable had happened; Route-masters belonging to another operator were put up for sale. Observers of the London bus scene waited with interest to see what London Transport would do. Earlier in the decade another very specialised type of bus – the Guy Wulfrunian – had come on the market and its main protagonist, West Riding, had sought to purchase as many as it could; would London Transport do the same with the Routemaster? At the time 328 of the 2875 Routemasters built were outside London Transport ownership. London Country had 209, British Airways 65 and Northern General 51, in addition to which RM 1 was with Lockheed, RML 2691 with Gala Cosmetics, and RM 3 was privately preserved.

The vehicles in question were thirteen coaches surplus to the requirements of British Airways who notified their intention of disposing of them in the summer of 1975. London Transport was, indeed, interested in acquiring them. Faced with a desperate shortage of serviceable rolling stock through inability to cope with the mechanical eccentricities of the now huge fleet of modern rear-engined vehicles, and beset with a horrendous spare parts shortage, the Executive was forced to take

desperate measures. It had scoured the world for spares and was even now just receiving a huge consignment of Routemaster piston rings from Federal Mogul of the USA; Leyland PD3s were being hired from Southend Transport; now the purchase of thirteen additional Routemasters presented itself as another small but positive way towards a solution. Though the front-entrance airport Routemasters were different in many ways from the standard London model and would clearly not be suitable for heavy in-town work, there was perceived to be a role for them in the suburbs and it was determined that, if purchased, they would operate from North Street garage on Romford area route 175.

British Airways' traditional sales policy had been to seek competitive prices for surplus vehicles either through tender or auction and the Authority decided to pursue this policy with the Routemasters. This left London Transport in the quandry of being unaware what the vehicles would cost, but in its determination to acquire them a sum equating roughly to £7,000 per bus was set aside as the price that it was judged they might fetch at auction. At the same time, however, heavy pressure was put on the airline to sell direct, to

which they finally acquiesced, enabling the thirteen vehicles to change hands on 29th August 1975 for just £3000 a piece. They received fleet numbers RMA 1–13 in the order of their registration numbers but – in common with recent types of vehicle such as the BL class of Bristol LHs and the MD class Metropolitans – they were not allocated body numbers.

On 11th October 1975 the RMAs entered service on the 175 replacing RTs. Being the airport coaches with the oldest certificates of fitness, they were all in shabby exterior condition, but such was the current need for them that there was not time to replace their tangerine and white livery or their British Airways advertising slogans such as 'Pound-stretchers to Paris'. Certain modifications had been made at Chiswick tram depot to suit them for their new role including removal of the luggage trailer tow bar and internal baggage racks and the addition of bells and used ticket boxes. No attempt was made to fit proper destination screens, and instead a black-on-white slipboard was carried in the nearside bulkhead window as the sole and totally inadequate means of route display. White fleet numbers were applied using regis-

Facing Page **RMA 9 waits outside North Street garage in January 1976. The usual London Transport insignia are present but retention of the well-worn orange and white livery is a sure sign of the desperate measures needed to augment the Company's serious deficiency of serviceable rolling stock as quickly as possible, and the vehicle's grubby nature is heightened by adverts for a 'girlie' magazine. The route destination slip board, which was barely adequate by any standards, could hardly be read after dark.** Brian Speller

Left **A few months earlier, soon after entering service on the 175, RMA 12 looks more like a mobile advertisement for British Airways than a London bus.** Capital Transport

tration letter transfers. A programme of re-certification and repainting into fleet livery commenced with RMA 1 which appeared as the first red RMA on 18th December 1975; the last to run in gaudy orange did so on 1st September 1976. This was RMA 8 which, along with RMAs 7 and 11, did not re-enter passenger service after being repainted red. During recertification a conventional illuminated route number box was mounted under the canopy. It had also been the intention to fit an illuminated bulkhead window blind display in place of the slipboards, but although the blinds were manufactured, this plan was dropped.

The continued operation of the RMAs was now, in any case, problematical. All along they had met resistance from the trade union on the grounds that the lack of internal stanchions and external destination blind displays made them hazardous and difficult to operate. They were gradually withdrawn from service during the middle months of 1976, the last ones, RMAs 3, 5, 6, 8-10 and 13, being delicensed on 2nd September.

Undeterred, London Transport had decided to convert the RMAs into driver training buses, there being an urgent need to replace the large number of RTs still attached to the Chiswick training school. This was no simple task as the whole of the staircase had to be removed in order to position the instructor behind his trainee, with a conventional window inserted in the staircase panel. A seat from upstairs was placed for the instructor in the vacant space. The bulkhead was modified to allow the instructor greater access to the driver's cabin, but because of the impossibility on a Routemaster of the instructor easily reaching the handbrake in an emergency because of its positioning, an auxiliary air brake had to be installed in the form of a handle located on the front bulkhead and actuating on the rear wheels. The semi-automatic mode of gearchange with which the vehicles had operated at North Street was retained in their training role even though the main Routemaster fleet for which the drivers were being trained consisted entirely of fully automatic vehicles. The driver's sun visor was removed and his small nearside driving mirror was replaced by a large RM type. An instruction was issued that the paraffin heater was not to be used.

Rumour had it that London Transport did not wish to incur the expense of fitting a full indicator display when the RMAs were taken into Aldenham for repainting, but some of the hostility to the vehicles might have been alleviated by doing so and the canopy number box was barely adequate as a compromise. **RMA 13 sports its new red livery at Chase Cross.** David Stuttard

Conversion into training vehicles was carried out at the Chiswick experimental shop to which RMA 4 was despatched in early October 1976, followed a month later by RMA 3 and RMA 5. Progress was slower than was hoped with the result that RMA 4 was not ready to commence its new role until 1st February 1977 and RMA 5 on 17th of the same month, both at West Ham; RMA 3, plagued by engine trouble, was not available until 20th June. At that point enthusiasm ran out and the remaining ten vehicles remained in store throughout 1977. Meanwhile a second batch of RMAs had come on the market and were acquired by direct negotiation on 9th November 1976 with the express purpose of using all fourteen as trainers. A sum of money was authorised to convert all 27 RMAs, including the work carried out on RMA 3–5 which were, in fact, the only ones actually dealt with under this authorisation. Also included was the cost of removing and blanking over the upstairs quarter drop windows for use elsewhere in the fleet, a task which was never started because spare windows later became available from another source – London Country. Like most of the first batch, RMAs 14–27 remained in store throughout 1977, and by the time they were revitalised in 1978 the secondhand Routemaster position had been transformed and they were no longer needed as trainers.

Rumours had circulated since the latter part of 1976 that London Country was soon to commence disposing of Routemasters. In due course London Transport was sounded out and proclaimed its willingness to consider buying any vehicles which may be available. Despite the temporary setback with the RMAs, currently nearly all delicensed, the Executive was anxious to consider any Routemasters made available from whatever source and was especially eager to get its hands on London Country's RMLs. Even the RCLs, if available, could possibly be modified for passenger service within London, and whilst the low capacity RMCs were discounted for this purpose they could be useful as staff buses or, preferably, trainers. The idea was now formulating that the RMAs could become staff buses, very little conversion work being required, with the rear-entrance RMC being a cheaper option for training work. Mindful that London Country possessed 97 RMLs, acquisition of these would materially ease an embarrassing situation which had arisen with Fleetline overhauls. Aldenham had been unable to cope with these up to the required standard, and even a lesser overhaul with consequently shorter certificate of fitness was still failing to keep the DMS programme on time. By purchasing the RMLs an equivalent number of DMS overhauls, at a cost of £10,000 each, could be avoided and the buses sold.

In October 1977 London Country produced its first list of Routemasters for disposal comprising 38 vehicles (29 RML, 2 RCL and 7 RMC) which were described as unserviceable. London Transport agreed to purchase these virtually at scrap value but also came to an agreement in December 1977 to buy an additional 30 regarded as runners. Within a very short space of time all 68 vehicles were inspected and their future determined; on 22nd December the first five (RMCs 1489, 1491, 1494, 1497, 1502) passed into London

Transport stock and by the end of the year 38 RMLs, 10 RCLs and 20 RMCs had all changed hands. Although London Country had considered 38 to be permanent non-runners, London Transport took the more optimistic view that 17 of them could be saved; the remaining 21, of which no fewer than 17 were RMLs, went straight to Wombwell Diesels' scrapyard without ever reaching London Transport property. At Barnsley all useful body and mechanical spares were salvaged and returned to London Transport to bring welcome easement of its chronic spare parts shortage.

January saw a minor flood of ex-LCBS vehicles entering service in one capacity or another. A particularly welcome addition to the operational fleet during the month comprised nine RMLs, all hastily repainted red at Aldenham. The first, RML 2347, was licensed only as a trainer but the remaining eight (RMLs 2315, 2332, 2351, 2419, 2435, 2444, 2454, 2460) were destined for passenger service. Also licensed in January but still in green livery for training duties were eight RCLs and thirteen RMCs. In February a start was made in phasing the RMAs on to staff bus duties. At first only vehicles from the second batch were used, still in British Airways' blue and white, but red liveried RMA 1 was reactivated as a staff bus in April and others of the earlier series followed suit later in the year. A programme of overhauling the ex-LCBS RMLs was instituted, placing a good deal of pressure on Aldenham's resources as all of them required a minimum of two months in the shops making good arrears of maintenance and replacing cannibalised parts. The first fully overhauled RML to take the road

Despite the fleetname, RCL 2219 was actually a trainer in London Transport ownership attached to Mortlake garage when photographed in July 1979, eight months after starting back with the old firm. At the time most garages were so preoccupied with maintaining their scheduled service that training vehicles tended to take low priority, but the retention of a former operator's name for such a long period on this and a few other vehicles was very surprising. D.W.K. Jones

RMC 1464 was a June 1979 acquisition which languished out of use until April 1980 when pressed into use as a Sutton based trainer. Though rather more tidy in appearance than many of its contemporaries, some of which retained green livery for several years, RMC 1464 was repainted red in February 1981.

was RML 2317 at Hendon on 13th March and during the year a further twelve followed, including one of the nine which had been repainted only a few months earlier, RML 2435. By the end of 1978 25 RMLs, 19 RCLs, 18 RMCs and 22 RMAs were operational, this total of 84 secondhand buses proving that the acquisition policy was already beginning to take effect. Two RCLs were of particular interest. RCL 2232 appeared repainted in standard red livery early in December 1978, the only one of its class ever to do so whilst on training duties. RCL 2221 had already been withdrawn after a mere fortnight as a trainer and was being converted at the experimental shop into a mobile cinema with 28 seats on the upper deck behind a large screen on a bulkhead installed at the rear of the first bay. Display panels occupied the lower deck and the vehicle was carpeted and fitted with additional lighting in both saloons. It was duly repainted in the strikingly handsome green, cream and red 'Shillibeer' livery chosen for the celebration of 1979 and made its public debut at the Easter Parade in Battersea Park along with a fleet of standard RM 'Shillibeers'.

Meanwhile, in an attempt to obtain additional bodywork spares London Transport had tapped another source when, in January 1978, it purchased two Routemasters from Northern General. This operator had recently commenced withdrawal of its Routemasters and, though it had dismantled a few to keep its own fleet going, others were sold to dealers. London Transport's two, which were NGT Nos. 3074 (RCN690) and 3101 (FPT587C), also passed in fact straight to a dealer, Wombwell's, where the required spares were reclaimed prior to the sale of the hulks in May 1978. These two vehicles never passed into London Transport stock or allocated fleet numbers, but in a way they were precursors of what was to come.

Also in January 1978, agreement was reached for London Transport to purchase a further 96 London Country Routemasters as and when they became available, an event which was thought to be fairly imminent as the Company was hopeful of divesting itself of the entire Routemaster fleet by the end of the year. This time everything went wrong. London Country had wanted London Transport to commit itself to take all 140 remaining vehicles (excluding RMC 4); London Transport wanted London Country to make the RMLs available in advance of the other two types which, in the absence of an agreement to purchase its entire stock, London Country would not agree to. The contract specified that London Country would bring all vehicles to Certificate of Fitness or at least MoT standard before handing them over, for which London Transport agreed to supply the necessary spare parts but then failed in many instances to do so. London Country was soon complaining that its pit space allocation for this work was under-utilised because parts were not available, and that mechanics had been allocated to do work which could not be done. London Transport, agreeing that it had failed to honour its commitment over spares, suspected that London Country, with its serious NBA (No Bus Available) problem, could not have done the work anyway. Soon delicensed

Routemasters were lying idle all around the fleet and London Country was faced with requests from its own staff to move them so that day to day work could be carried out. Three RCLs were handed over to London Transport in March 1978, and the Company was unable thereafter to get any more completed until August; by the end of 1978 only 18 out of the 96 had gone. Both parties were very dissatisfied with the deal and its almost total lack of progress.

An event which precipitated action on the part of London Transport when they heard about it was the sale in March 1979 by London Country of seventeen Routemasters to Wombwell Diesels for scrap. London Transport, who had hoped all along to buy the entire LCBS stock but declined to commit itself to a deal, made immediate contact with Wombwell with a request to take over the purchase contract for the seventeen buses. The dealer proved willing to negotiate and in due course the Routemasters passed into London Transport ownership, their place at Wombwell being taken by an equivalent number of SM family AEC Swifts. Fortunately only two vehicles (RMLs 2423 and 2424) had actually arrived in Barnsley where many of their external panels were immediately removed; the remainder were still on London Country premises. Galvanised into a deal, London Transport now agreed to buy all of the Company's remaining Routemasters. A much more realistic approach provided that the vehicles would in future be handed over in whatever condition they happened to be in at the time, the onus on making them serviceable falling upon London Transport. This meant a renegotiation downwards of the purchase price for the vehicles already contracted and was accompanied by an agreement that not fewer than sixty would be delivered by LCBS to LT designated storage sites by 30th June 1979 with the balance to follow by the end of the year. Furthermore London Country agreed to use its best endeavours to provide as many RMLs as possible in the first batch of sixty. They also agreed to house the remaining fifteen ex-Wombwell vehicles (RCLs 2220, 2226, 2231, 2241, 2243, 2260; RMC 1460, 1472, 1520; RML 2345, 2416, 2418, 2429, 2431, 2432) until London Transport could take them.

Both companies were experiencing severe capacity problems which had not been helped by the withdrawal by both of large numbers of rear-engined AEC single deckers. London Transport was faced with the need for covered accommodation for the many Routemasters which it intended eventually to restore to service but which would need secure storage meanwhile. The solution was found in June 1979 at the Royal Albert Dock in the form of a large modern steel and brick warehouse with a capacity for eighty buses. A licence was obtained from the Port of London Authority to occupy the building (officially known as Shed No. 2) on condition that buses would be parked in maximum groups of sixteen with 16ft fire breaks between each. From Monday 25th June London Country staff began delivering Routemasters to their temporary dockland home where each was topped up with two gallons of anti-freeze in the expectation that some may have to be stored for up to three years.

1979 proved to be the peak year for acquisitions. British Airways had ceased its Gloucester Road to Heathrow service on 31st March, rendering its 38 surviving Routemasters redundant. London Transport decided to purchase all of them if possible even though only 17 were actually needed for works transport. The remaining 21 could, perhaps, be converted into training vehicles even though the availability of capacity to carry out the conversions was in doubt. After once again overcoming British Airways' reluctance to conclude non-competitive sales, agreement was reached to purchase all 38 – this time at a bargain basement price of £2000 apiece – although the deal was not concluded until 5th June. Some of the RMA 28–65 batch were pressed into service almost immediately as staff buses in their blue and white colours but nine remained unused at Heathrow until September before moving to LT premises.

Skeletons at Wombwell. An appropriately gloomy day finds RMLs 2424 (left) and 2423 in the Barnsley scrapyard. These were the two Routemasters which London Country consigned direct to the breakers in 1979, causing London Transport to speed up its purchase of the remainder. John Fozard

Concurrently with the RMA negotiations, London Transport was considering the purchase of seven ex-Northern General Routemasters which had been put up for sale by E.H.Brakell (dealer) who, it was thought, had experienced difficulty in obtaining spare parts for them. The view was taken that they might be usable in passenger service and should be acquired, authority for their purchase being approved at the same time as that for the RMAs. The prevailing mood was now that any Routemasters coming on the market should be bought, and the idea was even mooted of discussing with Northern General the purchase of all their remaining Routemasters, believed to be about 24 in number. In June 1979 authority was received to purchase a further five of the type, four direct from Northern General and one from the Grays dealer, Ensign Bus Co. Ltd. Logically the class designation RMF was applied, the batch RMF 2761–2772 being ex-Brakell (RMF 2761–67), Northern (RMF 2768, 2769, 2771, 2772) and Ensign (RMF 2770). Transfer of ownership was not recorded as officially taking place until December 1979

(RMF 2761–69) or January 1980 (RMF 2770) with the remaining two following in August and September 1980 respectively, although they began arriving at the docks for storage in July 1979. All twelve were originally stored there although RMFs 2768/69 were taken to Aldenham in January 1980 for general assessment as to their future, with RMF 2769 passing to Chiswick in March for closer examination.

On the assumption that the RMFs would be suitable for service operation, route 26 from Finchley garage was selected as the likely venue for six of them, with others probably to operate on the Round London Sightseeing Tour. In the event of further acquisitions, Fulwell's route 281 might prove another possibility. Local managements, fraught with daily vehicle problems, were none of them too keen to accept another type of bus which they regarded as non-standard, even though they were Routemasters. Examination revealed a number of tasks which would have to be carried out for stage carriage use. The lack of internal stanchions which had proved one of

the main drawbacks with the RMAs on route 175, was also a feature of the RMFs. Clearly these would have to be provided to gain union acceptance even though the vehicles had successfully run on Tyneside without them, and problems with the manufacture of fixing plates etc were envisaged. The bells, which consisted of a single strip on each deck almost inaccessible to short people, would need to be augmented. Even the blind winding gear, which was only accessible from the upper saloon, could prove a problem over conductor acceptance with a union rigidly opposed to adopting any new working features without additional payment. The seats on most if not all vehicles would need retrimming, many now being of leathercloth replacing the original moquette, and even this task was made more complicated for Aldenham's flow line system because the seat frames were of a non-standard type. A minor London requirement, the provision of a holder for the conductor's waybill, had been met on RMF 2769 when one was fitted on the staircase partition soon after the vehicle arrived at Aldenham. Mechanically the greatest worry lay with the differential which was totally unlike the London type and was rumoured to have been a major factor in Northern's decision to withdraw the vehicles, not because of any inherent unreliability – they were arguably better than the London model – but because of extreme difficulty in obtaining spares for this now-obsolete AEC design. As with London's RCLs, the Northern vehicles had been built without provision for conversion to fully automatic operation, but this was thought to pose no problem. The fitment of automatic rather than manual brake adjustors was considered essential.

By mid-1980 it had been decided to use the RMFs only on Round London sightseeing, any hope of putting them into stage carriage service having been abandoned. Enthusiasm for them was on the wane, and their run-down appearance in shabby, faded poppy red was not encouraging. When, in February 1980, Brakell offered three more, which had been running on hire to London Transport through Obsolete Fleet Ltd on the sightseeing tour, the offer was declined even though current certificates of fitness were held. Similarly the opportunity arose soon afterwards to acquire a further 17 direct from Northern (one immediately, six later in 1980 and the remainder in 1981); this too was declined. Though probably a wise decision, from an enthusiast's point of view it was a pity as it precluded the return to London of RMF 1254.

1980 proved to be a bumper year for getting the ex-LCBS RMLs overhauled and into service, only seven still being outstanding at the year end. Most retained their own bodies on overhaul, the only ones to become intermixed with the main fleet being RMLs 2332/2354. Even after a subsequent overhaul the great majority kept their original bodies and will now do so until the end of their lives. The very last to be put into service was RML 2345 at Bow in May 1981 after an overhaul which lasted for no less than six months. Reduced to little more than a shell through cannibalisation whilst parked in Grays garage, it required considerable determination to restore as a complete vehicle. Only four RMLs operated for London Transport in green livery, all as trainers, these being RMLs 2346, 2353, 2413, 2451 which were all pressed into service in September 1979 to counter the shortage of training vehicles. Although the first two lasted only a short time before going off to Aldenham for overhaul, the higher numbered ones were trainers for longer, RML 2413 even being put through an MoT test to survive as a trainer through to January 1981.

September 1980 marked the start of a programme for repainting the final batch of RMA staff buses from blue into red livery, though not all were dealt with and four years later RMA 50 was still working from Fulwell in blue. The scheme to convert 21 of them into trainers never came fully to fruition, only seven being dealt with between April and June 1981 (RMAs 29, 38, 40, 42, 47, 55, 60). These entered service in red livery and, together with RMAs 3–5, brought the RMA training fleet to its maximum of ten. The same month of September 1980 saw the first RMC trainers in red livery commencing with RMC 1469, although the original repaints involved unlicensed vehicles and the first red one to actually take to the road was Kingston trainer RMC 1475 on 21st November 1980. Thirty-two of the class began their training career in NBC green, sometimes initially still carrying London Country fleet names (as did some of the RCL trainers). A smaller number, 21, were repainted red before commencing training duties. Gradually London Transport features such as black mudguards and red-oxide coloured wheels began to embellish the green

vehicles whose ranks diminished considerably in 1980/81 through repainting. However some escaped the net; indeed RMC 1480 was even repainted in NBC green at Sutton garage in October 1980. This was the penultimate green one to survive, being outlasted for a month by RMC 1516, a North Street based vehicle which did not receive red livery until as late as May 1988, a full decade after returning to London Transport ownership. Most vehicles retained their rear air suspension although RMC 1518, which was employed on the famous Chiswick skid track, was converted to coil springs in August 1980 when the work was found to be too strenuous for the air bags, and RMC 1470 was also converted to coil in December 1981. The vehicles retained their coach-based rear axle ratios until replacements were needed.

Above **RML 2346 spent comparatively little time out of use. Still active at Windsor until as late as May 1979, it ran as a trainer with London Transport in green livery for a few weeks from September and, after overhaul, was in service in red livery by January 1980. It arrived at Cricklewood for route 16 in May of the same year.** Eamonn Kentell

When the RMCs were repainted red little if any tidying was carried out on the bodywork as witnessed by the dents in the front dome of RMC 1518. Repainted in October 1981 with old-style gold fleetnames, this vehicle had been fitted with coil springs for use as a Chiswick skid bus, a role shared with RT 1530, and the second view shows it at work as such.
Capital Transport/Colin Fradd

Below RMA 47 was one of the second batch of conversions to trainer requiring the fitting of a window in the panel behind which the staircase was formerly housed. Newly in service as a Clapton trainer after being repainted red in June 1981, it takes a break at Manor House. The same location finds Tottenham trainer RMA 55 cruising past in September 1987 with its original coat of red paint still looking fairly smart. All the RMA trainers had 'Driver Under Instruction' notices posted in the blank space where destination equipment should have been.
Joel Kosminsky
R.J. Waterhouse

The RCLs looked particularly splendid after their very thorough Aldenham overhaul into red livery and remained free of advertisements much longer than normal. RCL 2223 was one of those to operate from the first day. Even from a distance the RCLs were easily distinguishable from other red Routemasters by the nearside positioning of their route numbers. Capital Transport

Below Right **Conversion of the RCLs to their central bus role included removal of the platform doors and the fitting of new grab rails and central stanchion as demonstrated by Edmonton's RCL 2259. The only variation from this was on RCL 2256, a Stamford Hill bus, which was fitted with a standard Routemaster rear end after a serious accident in April 1981.** Alan B. Cross

The biggest transformation was with the RCLs. Although 27 out of the 41 had been employed initially as trainers, it was decided to place all of them except cinema bus RCL 2221 into public service to speed up the removal of the DM class from crew work. Eight had even been converted into permanent trainers with auxiliary air brakes etc but it was decided to deconvert them. In January 1980 RCL 2239 was taken out of storage at Royal Albert Dock for a pilot overhaul at Aldenham. Included in the schedule of work was the removal of doors, disconnection of wiring and plating over of switches; removal of luggage racks and upper saloon ash trays; renewal of the platform covering and provision of a stanchion; and conversion from double to single head lamp. The vehicle was completely retrimmed with blue moquette as used on the DMSs and other types, with dark burgundy rexine seat backs. The deep, comfortable squabs and cushions were retained, and the overall interior effect with a combination of blue seats, red panelling, yellow window surrounds and white ceilings was very colourful and bright. The remaining vehicles followed RCL 2239 through Aldenham between February and September 1980 to produce a fleet of what must surely be the

In April 1981 the exhibition and cinema bus, RCL 2221, lost its original Shillibeer livery in favour of a red and yellow combination not dissimilar to the style in which RCL 2260 was latterly painted. It is seen outside the Covent Garden museum in June 1982. *Alan B. Cross*

most attractive and comfortable stage carriage vehicles ever to grace London streets. All retained their air suspension, and though there was no deliberate policy to replace the AV690 engines with the less powerful AV590 many were, in fact, replaced as a matter of necessity. After an initial suggestion that the RCLs might operate from Holloway and Walworth garages on route 45, they finally settled at Edmonton and Stamford Hill to become a daily feature on route 149 and, though not scheduled to do so, were frequent performers on all other RM routes from these two garages. The first ones entered service at Stamford Hill on 10th August 1980 with Edmonton's allocation following on through to October.

By 1981 future fleet requirements could be envisaged with much greater clarity than had been the case earlier, and it was perceived that, far from facing a shortage of Routemasters, a quantity of standard ones would almost certainly become surplus in 1982. Inevitably thoughts turned to disposing of the non-standard types, even the RCLs which had not long since been placed into service, but with first priority on the fair number of RMA, RMC and RMF types which had still not been found employment or which, in a few cases, had broken down after a short while and been

virtually abandoned. The RMFs, in particular, were undisputed candidates for disposal, doubts having been expressed for some time over the wisdom of having purchased them in the first place. All the disused vehicles were collected into Aldenham for inspection, and disposals commenced in June 1981. In all, 36 unused vehicles were sold comprising 13 RMCs (RMCs 1454, 1455, 1460, 1463, 1465, 1472, 1478, 1479, 1482, 1487, 1493, 1514, 1520), eleven RMAs (RMAs 17, 21, 28, 35, 39, 43, 45, 49, 54, 56, 57) and all twelve RMFs. Two of the RMCs (1455, 1493) were, in fact, sent for scrap from Aldenham in error, being serviceable vehicles which had been sent there for training handbrakes etc to be fitted. With one exception all 36 were sold in their pre-LT liveries, the odd man out being RMC 1487 which was repainted red and then declared beyond economic repair. In addition to these vehicles one RMC and four RMAs which had been used for a while but had become unserviceable were also disposed of. The majority were scrapped but a few went to preservationists, two RMFs were sold to become mobile restaurants, whilst RMA 56 stayed close to home by being sold to the sports club at Plumstead garage.

Glory for the RCLs was short lived. They performed well and were popular, but the quest to standardise led to their demise in 1984 after serving the Hertford Road for four years. Anticipated problems with their air suspension had proved largely unfounded, their main source of trouble having been wearing-out of the gangway floor coverings,

the flat coach-type mat being inadequate to withstand the heavy pounding on the 149. Many had to have replacement, slatted floors fitted which was a time-consuming task. Two RCLs were withdrawn in late 1982 and seven more in 1983; a big run-down in 1984 resulted in the last Stamford Hill vehicle (RCL 2222) running on 1st December and the final Edmonton one (RCL 2260) on the 15th of the same month. All were put into storage and several were disposed of, but fortunately not all, and in 1986 the eleven which remained were placed back into service on London sightseeing as recounted in Chapter 14. During 1982/3 the authorities in charge of the London Transport Museum decided to preserve an RCL as the final representative of Green Line development during the London Transport era and selected the last of the class, RCL 2260. However on later examination, RCL 2249 was deemed to be more suitable and passed to the museum in March 1985. RCL 2221, the cinema bus, continued in existence, being repainted from Shillibeer colours into red with yellow central band and window surrounds in 1981, and further repainted in all-red except for a cream band in 1987. Regarded as a unit of the fleet of miscellaneous service vehicles though retaining its bus stock number, it passed on 2nd January 1986 into the control of Distribution Services, a semi-autonomous profit centre within London Regional Transport. During the summer of 1989 it toured Britain in white and green livery under the government sponsored Motability campaign for the use of lead-free fuels.

With the gradual demise and final closure of Aldenham the requirement for a staff bus fleet diminished. When LRT Bus Engineering Ltd (better known as BEL) was formed as a separate business under the London Regional Transport umbrella on 1st April 1985, 34 RMAs were included within its assets, many of which were surplus to requirements. Most were sorely in need of repainting and in about June 1985 BEL unveiled a new livery on RMA 16; a drab all-over dark grey with black wings and silver wheels. No names or fleet numbers were carried. The gloomy effect was lightened somewhat on later repaints which were given red window surrounds and carried the BEL symbol between decks, but only RMAs 6, 8, 9, 19, 62 were dealt with; the remainder stayed red. The fleet represented an altogether bad image for a company specialising in vehicle painting and repair in the run up to privatisation, which came in due course with its sale in January 1988 to Frontsource Ltd. During 1987 many surplus RMAs were sold and in December the last two, RMAs 8 and 16, were withdrawn, RMA 16 making the final journey on the 17th to St Albans. Such staff as still needed transportation within this much contracted business could be accommodated in minibuses. At the end of 1986 six RMAs surplus to BEL's needs were transferred back into London Buses stock for sightseeing work with its Commerical Operations Unit as described in Chapter 14; the same unit also acquired RMC 1491 in July 1987 which was repainted for its continued use as a trainer into a red version of the traditional Green Line livery with cream band and window surrounds, red (as distinct from red oxide) wheels and wings.

Top **Edmonton garage staff adopted RCL 2260, the highest numbered of the class, as their Showbus, embellishing it with cream band and window surrounds, gold transfers, and restoring the original twin headlight layout. In this guise it looked very attractive indeed. It is seen at Seven Sisters in April 1984.** Colin Fradd

Centre **As part of its relocation from Aldenham, BEL purchased a brand new spray paint booth for installation at Chiswick capable of completing eight vehicle coats a day. What better vehicle to publicise it than one of the Company's own staff buses, RMA 9, which is seen in the final stages of receiving its uninspired new grey livery. RMAs 9 and 59 from 1984 onwards carried DMS-type cant-rail flashers in place of the normal 'ears'. There may well have been others.** BEL

Left **RMA 8 passes Acton Town station shortly after leaving Chiswick Works with staff on their way home after the day's activities. Red window surrounds provide some relief to the grey livery.** J.H. Blake

Facing Page Top **RMC 1515's 1987 conversion back to PSV standard by Forest District as an open topper was a welcome move but its employment thereafter was spasmodic. In December 1988 it stands at Marble Arch on a short working of route 15 for the benefit of those wishing to obtain a better view of the Oxford Street and Regent Street Christmas lights.** R.J. Waterhouse

The RMCs, though prominent on London streets for many years in their lowly training role, led an uneventful existence until 1987 when RMC 1515 took on a completely new role as an open-topper. Under the initiative of London Buses' erstwhile Forest district, it was converted for its new role at Leyton garage and appeared in red livery with green band and upper parts as the 'Forest Ranger', available for hire work and for special operations such as the special Christmas Lights extras on route 15. RMC 1515 was first licensed in its new guise on 1st October 1987. Almost simultaneously RMC 1469 was being converted by Cardinal district into a specialised recruitment and training vehicle with reception, interview and test rooms and fitted out for mains heating and lighting. As originally outshopped in December 1987 it sported a striking livery of red with yellow lower saloon window surrounds and central band and silver wheels, but by the spring of 1988 this had given way to standard livery in deference to the Company's official policy on livery matters. This did not deter the new East London unit from adopting gold central and window reliefs, plus gold lettering, when 'The Beckton Express' route X15 began on 15th March 1989 using refurbished RMCs.

The X15 was conceived as a high quality commuter service and it took a high degree of commercial courage and optimism to convert a small fleet of run-down, 25-year old ex-trainers with which to operate it. However, six RMCs (1456, 1458, 1461, 1490, 1496, 1513) were hastily but very well renovated by engineering staff at Upton Park in a remarkable transformation which brought the class into regular central London passenger carrying service bearing red livery for the first time. The X15, which was unique among a plethora of London commuter services in employing conductors, started with five journeys in each peak, inwards from Beckton to Aldwych in the morning and reversed in the evening, but this was increased to six from 20th March and the service extended to Oxford Circus. The six RMCs were joined by a seventh, RMC 1485.

On conversion for their new role all remaining luggage racks were removed from the vehicles (many already having been taken out during their time as trainers) and seats were reupholstered in standard RM red moquette. RM-type front indicators were fitted and the gearboxes were converted to fully automatic. The impressive red and gold livery also saw a welcome reversion to gold fleet numbers, and the East London unit name and Thames sailing barge logo were also in gold.

The large-scale Routemaster acquisitions during the height of activity in 1977–80 required a capital provision of £2.3m for purchasing and renovation, which was a large outlay for a fleet of second hand buses. Even though some were never used, the great majority gave excellent service in their new ownership and many continue to do so. The investment has paid for itself many times over. The vehicles have also added an extra dimension of interest to the London bus scene, and with many being strictly speaking non-standard, they have tended to adopt in a few cases unauthorised modifications which completely standardised types may not have done. A few of these modifications have included RCL 2256 which, after an accident in April 1981 appeared in July with a standard RM open rear platform; however as it was not at the same time provided with the lower saloon offside emergency window carried on the RMLs and therefore had no secondary means of emergency escape on the lower deck required for vehicles of its length, the legality of the conversion under the Construction and Use regulations remains in doubt. At some time in its training career – probably in 1985 – RMA 60 was rebuilt with a standard RM front end assembly with only single headlamps, as indeed were RMCs 1474 and 1494. RMC 1476 ran in latter years with the rear destination indicators removed.

The Beckton Express of March 1989 marked the biggest revival in fortunes for the RMC class after years of relative obscurity. The unusual red and gold livery graces RMC 1496 as it stops to pick up non-existent passengers in Duncannon Street. Loadings on the service did however become reasonably healthy, justifying the faith placed in the operation by the East London unit.
Ken Blacker

CHAPTER TWELVE
ALDENHAM WORKS

At the start of the 1970s each Routemaster was scheduled to visit Aldenham approximately every 3½ years on a cycle alternating between heavy overhaul and body repaint. A variation came in 1971 when part of the intake was given only a light overhaul, attracting a shorter length CoF. This process complicated the overhaul cycles with the result that some Routemasters had subsequently to be submitted for their fourth overhaul ahead of their contemporaries. Over the years, floats had been set up and dispersed many times to accommodate various types and batches of Routemaster. For instance, the early RMLs required their own float, as did the buses with illuminated exterior advertising panels, both RM and RML. The RMC class was overhauled once during London Transport ownership using the float system but when the BEA coaches were given their only overhaul it was on an individual basis with no change of identity. Their trailers were also overhauled at Aldenham.

Although it was still lauded as a showpiece, there was a slowly developing awareness that the Aldenham factory was overelaborate; a diminution of its work load was occurring and a question mark was beginning to appear over its long-term viability. Continuous reductions in the size of the fleet; the loss of the Country Bus and Coach department on 1st January 1970; the influx of large numbers of off-the-peg buses unsuited to Aldenham's flow-line methods, many of which did not last to see their first overhaul anyway; these were all factors which contributed to the run-down of the works. It was common knowledge that staff productivity was far too low and the quality of their output came increasingly into question. Closure was mooted more than once; in 1973 the GLC Transport Committee considered moving the whole operation to a much smaller site but decided that the financial advantage of doing so would be slight. Ten years later, with the future looking very

fragile, a review concluded that it would be difficult if not impossible to make the place financially viable and recommended its total closure. It was calculated that, by contracting work to outside companies, no less that £8 million pounds per annum could be saved whilst the sale of the site would realise a further £25 million. The decision to close was still not taken but, in the meantime, routine overhaul work was drying up.

The introduction, from 1st January 1982, of the EEC based system of annual vehicle inspection caused a major rethink for London Transport of its whole approach to maintenance, the first necessity being the reduction

Above **Aldenham, August 1978. RM 1029's third overhaul has just been completed and it is now almost ready to pass to the recertification shop. RML 2608 from Hendon and RML 2722 from Stockwell have both received a repaint; the spray booth maskings have still to be removed and the black paintwork and transfers added.**
John Hambley

Above **Overhauled vehicles occasionally missed the standardisation net and there were always two or three RMs in the fleet not fitted with the full width front band. One such example was Muswell Hill's RM 208, carrying late body B1907, in August 1970.** Colin Stannard

Below **At the fourth overhaul cycle for the earliest bodies a serious attempt was made to bring them back to fleet numbers below 1000, and had almost succeeded when RM 1691 emerged at Stamford Hill carrying body B243 in October 1977. A few more followed in 1981.** J.H. Blake

The widening of periods between overhaul, with intermediate repainting, resulted in patchwork vehicles such as this becoming fairly commonplace for a while. Dented panels were not replaced at Aldenham on vehicles which had merely come in for a repaint, it being the responsibility of garages to deal with them. Upton Park's RML 2588, photographed on 8th June 1976, is scheduled to go for repainting in two days time and has been repanelled in preparation for this. Alan B. Cross

By far the lowest numbered Routemaster to receive an illuminated offside advert body was RM 796 and it still carried this at the time body changes at overhaul came to an end. It rounds Hyde Park Corner in August 1984. Colin Stannard

With other more pressing problems on hand, garages began forgetting the preparation of vehicles for repainting, with the result that the dents with which they went to Aldenham returned back sprayed with fresh paint. Standards were visibly on the decline by March 1980 when Croydon's RM 1707 was photographed. Vehicles had been outshopped free of advertisements from about April 1977 onwards.
Alan B. Cross

of the rota system from a 64 week to a 48 week cycle, bringing it within the span of the annual Freedom From Defect test. It was clear that the new, far more onerous, arrangement would place greater emphasis on vehicle overhaul within garages, eight of which were set up as approved test centres in addition to Aldenham, and it was duly decided that each vehicle would undergo its test preparation at its home garage for three years, being dealt with at Aldenham in the fourth, where it would also be repainted. Until 1982, overhaul work at Aldenham continued much as before but change was in the air and in midyear the RM works float was officially disbanded, 19 years after it had been set up. The last vehicles to be resuscitated from the float, RMs 2011/2117, were licensed for service on 28th October and the last Routemaster to be outshopped under the old system was RML 2760 on 5th November 1982, now re-united with its original body. For a while past, the tendency had been to relink bodies with their original fleet number on overhaul where possible though there appeared to be no particularly good reason for this. The latter part of 1983 saw trials on the overhauling of Routemasters without dismounting bodies and the final RM body lifted at Aldenham was reputedly RM 198 (body B155) on 28th June 1984. With no float in existence, the exchanging of identities diminished and ceased completely in March 1984 with the exception that, in October 1984, RMs 6 and 192 exchanged bodies during overhaul in order to restore body B6 to its original fleet number. At about the same time, a batch of fifty vehicles including twenty Route-

masters was despatched to outside contractors for trial overhauls, a cost cutting exercise which temporarily heartened Aldenham staff when they learned that certain vehicles were away from London for excessive periods of time, some for up to eight months, and even then did not always display workmanship of an acceptable standard.

The cessation of number changing in 1984 enabled a stable situation to be achieved for the first time in regard to bodies with non-standard features such as non-opening front windows and offside illuminated advertisement panels. With regard to the latter, whilst some attempt had been made to keep them within their original batches, the final position was that eleven RMs and five RMLs carried fleet numbers not originally associated with illuminated advert buses. Final fleet numbers were: RM 796, 1528, 1849, 1906, 1916, 1924-1926, 1928-1934, 1936, 1939-1947, 1949-1956, 1958-1967, 1969-2004, 2006-2019, 2021-2051, 2053-2127; RML 2527, 2544, 2561-2567, 2569-2576, 2578, 2581-2647, 2649-2660, 2662, 2664, 2759.

Aldenham was restructured in a last ditch attempt to make it pay, with only sixty percent of the staff who had been there at the end of 1983 still on books in August 1984. As can so easily happen with large scale cutbacks, many of the older and more experienced staff accepted redundancy terms, leaving behind a demotivated workforce few of whom believed in the management's ability to keep the works alive. From 1st April 1985, the works had been separated from the operating wing of the bus business, being controlled by LRT Bus

Engineering Ltd, a wholly owned subsidiary of London Regional Transport, but even this new commercial approach had failed to stem complaints of poor quality control and overcharging which saw the cost of a Routemaster repaint some forty per cent more expensive than could be achieved outside, and likely to take twice as long. In its new commercial guise, Aldenham continued to deal with Routemasters and although useful contracts were obtained, including the renovation, fitment of public address equipment and, in some cases conversion to open top, of 50 RMs and RCLs for the Round London Sightseeing Tour, overhaul work had diminished and then ceased entirely. The last Routemaster to enter the process was RML 2376 on 27th November 1985 but its overhaul was not completed, in common with a number of vehicles, until March 1986. The cause of overhaul work drying up was a decision by London Buses Ltd in November 1985 to change its maintenance schedules once more, removing totally the need for vehicles to visit Aldenham. With a loss of £5 million per annum still persisting, and a once only cost of £2 million likely to be incurred in relocating to Chiswick such work as still remained, the announcement in June 1986 that Aldenham was to close came as no surprise. Its last day in business was 14th November 1986 and the works closed a mere shadow of the granduer that had existed thirty years earlier. Some of its last work was the repainting of Routemasters sold to Scottish Bus Group operators. Routemasters were embarking on a countrywide renaissance but for Aldenham works it was the end of the road.

Perhaps almost rivalling Chiswick's skid patch for fame, Aldenham's tilt test was another casualty of the drive for economy that saw both establishments close down in the mid-1980s. This public display was part of the open day held in 1979 as one of the Shillibeer anniversary events.
Capital Transport

The last job undertaken by Aldenham for London Buses Ltd was the conversion of a number of RMs to open-top and refurbishment of other RMs and some RCLs for the London Coaches unit. Still bearing the garage code of its last operational garage, Westbourne Park, RM 49 is decapitated at Aldenham in preparation for its new career as a sightseeing bus. A team of some 25 craftsmen and assistants was allocated by BEL to the preparation of the original fifty-strong sightseeing fleet.
LT News

Above Left **The despatch of 50 buses, including 20 RMs, for outside overhaul between January and March 1984 indicated that the writing was on the wall for Aldenham. Mechanical overhaul was carried out by Leyland at Nottingham and body refurbishment by Eastern Coach Works at Lowestoft. Outside the famous frontage of the ECW works, sadly now demolished, is a vehicle whose identity will be revealed as RM 1103 as soon as all the appropriate transfers become available!** Howard Smith

Above Right **Conditions in this ECW workshop were almost primitive compared with the flow line production methods which distinguished Aldenham in its hey-day. However they were probably far more cost effective, and not so unlike the paint shops of the various contractors to which Routemasters were despatched in subsequent times. Accompanying RM 1544 in the works where CRL 4 was built many years earlier, are new Leyland Olympians under construction for NBC subsidiaries.** Howard Smith

Centre and Left **In addition to the work on Routemasters for London's sightseeing fleet, other tasks undertaken at Aldenham during its last year included the repainting of RMs for Scottish Bus Group companies. In March 1986 a London Coaches RM is flanked by one at the start of the complicated paint job for Kelvin and one for Clydeside. The line-up of RMs at Southall for the Clydeside company was photographed in August 1986, three months before Aldenham's closure.** John Laker

CHAPTER THIRTEEN

LONDON BUSES LTD

On Friday 29th June 1984 London's busmen found themselves working for London Regional Transport. Outward signs of change were not immediately visible although in due course the legal ownership panels on buses began to reflect the change of name. The Secretary of State's intention to slash revenue subsidy for London's bus and rail services by more than half by 1987/8 did not augur well, nor did his intention that – as required by the terms of its Act – LRT should soon commence offering the operation of bus services for competitive tender in order to achieve a satisfactory level of service at the lowest possible cost. Also apparent was the fact that direct operation of bus services by LRT was very much an interim measure and that a separate company would be formed to take these over. Meanwhile, in October 1984, the bus business announced the measures it considered necessary to meet the harsh financial realities ahead which included an increase from 53% to 75% in driver-only operation by March 1988 and a reduction in staff (excluding any whose jobs may be lost through tendering) of 3,800 over the same period. As a result of the speeding up of opo conversion (as one-manning was now referred to) a further 300 Routemasters would be lost in 1985 alone.

Meanwhile the displacement of RMs from routes earmarked for driver-only operation pressed ahead. The New Cross share of route 141 and the Catford portion of the 180 both commenced progressive conversion to Titans on 25th July and were complete by mid-August – although RM sightings continued while crew operation remained. From 25th August Catford became a recipient of Ts for route 47 – although once again RMs continued to make occasional appearances – followed on 28th September by Bromley for the same route. Mid-October found D-class dual-purpose Fleetlines working as crew buses at Croydon on route 190 in place of RMs, whilst in the final months of the year the now ever more frequent mixing of types found Brixton running a jumble of RM, D and new M on routes 109 and 133 with Stockwell scheduled a miscellany of RM, RML and D on routes 77A and 88, these being joined by the unusual Ailsa Volvo V 3 from March 1985. On 27th October the penultimate crew-worked night service was converted to opo; this was Holloway's N93 which had been noteworthy since about mid-July in being worked by RMLs which looked strangely out of place amongst the host of rear-engined vehicles which made up the rest of the now much-enhanced night network.

On 10th October LRT put the first batch of services out to tender; twelve out of the thirteen were ones it operated itself, and covered 52 scheduled buses from twelve garages. However, RMs were excluded, as it was announced that no crew routes would be tendered. On 27th of the same month the new Norwood garage opened and Streatham closed on the same date for rebuilding; its staff transferring temporarily to Clapham whose Routemasters now adopted the garage code AK instead of N. A serious fire in Oxford Circus tube station on 23rd November caused the closure for a few weeks of the Victoria Line tube over its central section requiring many duplicates on Red Arrow route 500 to cope with commuter crowds to and from Victoria station. Most of the duplicates were Metrobuses working with conductors, but history was made on 27th November when RM 2068 worked a stint on the 500 in the first instance of a standard RM working on an opo service. This was followed on Saturday 23rd February 1985 by RM 595 reportedly working on opo route 59 from Thornton Heath garage, but this was now the era when the rigidity and certainty of vehicle workings, inherited from the LGOC and so long practised by London Transport, was rapidly breaking down.

Facing Page **Having lost control of London Transport in June 1984 the Greater London Council found itself fighting for its very life against a government determined that it should not survive. Three weeks into the new era Croydon's Showbus, RM 1000, still carries the exhortation to keep the GLC working for London. The fine condition of this vehicle, with many original features restored, typifies the keenness of local garage groups soon to be discouraged by a management change of attitude.** Alan B. Cross

This Page **In October the venerable if antiquated Streatham garage closed so that a (short-lived) new phoenix could arise on its site. To accommodate Streatham's vehicles while rebuilding was in progress, the old bus garage at Clapham continued in temporary use. At the same time the new Norwood garage came to life and former Showbus RM 719 is seen attending the reopening ceremony. The non-mandatory 'No Smoking' window stickers on the upper saloon of RM 719 are a recent innovation.**
Ken Blacker/J.H. Blake/John Parkin

Fresh visual changes were made to a few members of the RM fleet in 1984 both of which were later to become widespread. First was the introduction of a new, flatter type of radiator grille on which the mesh was not so deeply inset and the top corner retaining brackets were almost flush with the polished surround. Second was the fitting of aluminium frames to the side and rear lower advertisement positions on RMs 1126, 1352 and a number of Metrobuses at Shepherds Bush. This was a trial run of a scheme proposed for the whole fleet which had been brought about by the need to overcome the paint-destroying effects of pulling off vinyl advertisements, whilst at the same time providing what in effect were mobile hoardings capable of attracting large scale advertising campaigns which could be implemented far quicker than with conventional posting.

1984 ended on a sad note with the withdrawal in December of the last RCLs from route 149 and the various other north London routes on which they had made their mark for comfort. The last one of all was Edmonton's RCL 2260 on Saturday 5th December. This particular vehicle had earlier been adopted as Edmonton's Showbus, and with the restoration of its twin headlights and the picking-out in cream of its window mouldings from Green Line days it looked particularly handsome. RCL 2260 was one of the vehicles which had not been restored to standard livery in response to management exhortations, which were repeated in 1985. On the unlikely pretext that the buses could cause confusion to passengers, management continued to insist that Showbuses must revert to ordinary livery, whilst acknowledging the help that had been given to the Company's prestige by staff who had prepared the vehicles and won many trophies. As a result some notable buses were bought by the staff when they became available for sale, such as RM 8 (whose RT style seat cushions had now passed to RM 86) and RM 1000.

After employing a standard range of sizes for advertisement panels over very many years, London Transport threw caution to the wind in the search for extra revenue and large untidy offside vinyl displays, which had first appeared in T configuration on DMSs, spread to RMs in the form of a fallen L. As in the case of Bromley's RM 263 which in July 1984 depicted someone climbing an offside staircase on behalf of Colt lager, these advertisements clashed unhappily with the lines of the bodywork. Furthermore they often left patchy areas of missing paint when removed. Prior to the introduction of advertising frames at the end of 1985, some RMs received enlarged nearside adverts as seen on RMs 501 and 1185 at Westminster. Colin Brown

The publication of LRT's annual business plan revealed that five bus garages were to close (Battersea, Edmonton, Poplar, Southall and Walworth) between November 1985 and an unspecified date in 1987. It was also announced that from 1st April 1985 LRT's bus operations would be vested in London Buses Ltd, a wholly owned subsidiary. London Buses had been the internal title of the bus division since 1979 when it was formed as an accountable business unit, but it would now become the public face of the organisation. London Buses Ltd was duly incorporated on 29th March 1985, a matter of days before it commenced trading.

Some 370 more RMs were disposed of during 1985, withdrawals coming thick and fast. A massive batch of one-manning on 2nd February mostly affected routes on which new-type buses had already been pre-positioned, but Tottenham's route 76 and Wood Green's share of route 141 had only seen the odd Metrobus or two and Holloway's route 4 had stayed solidly RM to the end. Despite the low morale which pervaded so much of the organisation, a strong affection lingered amongst staff for the Routemaster, heightened perhaps by the accelerating pace of withdrawal, and local endeavours at a number of garages saw RMs in use on the last day of crew operation on routes from which they had already disappeared. Thus Fulwell on route 33, Sidcup on route 161, Croydon on route 190 and Catford on route 208 all managed an RM on the last day, even though in some cases it meant borrowing from another garage. The last crew-operated night service, Wood Green's N29, also went one-man, and even here an RM was seen to operate on occasion during the last week where the class had never ventured before. Cricklewood's RMs commenced their departure from route 266 on 2nd February as Metrobuses became available, and the total withdrawal occurred of the ill-starred Kingston shoppers' services K1 and K2.

RM 1877 appeared in March 1985 carrying a livery of its own with white central wrap-around band proclaiming 'Advertising on the Move'. Allocated firstly to Tottenham, it transferred in August 1985 for a much longer sojourn at Clapham. The unique livery lasted until November 1987. The silver painted radiator grille was given to it at Clapham, along with a number of other RMs at this shed and at Norwood. Colin Brown

London Buses Ltd had come into being by the time of the next wave of changes on 27th April when Brixton's route 118 was one-manned with Metrobuses. This was the last day of crew work at Bromley who borrowed an RM to commemorate the event on the 47 and 119. On route 45, a short-lived Stockwell RM allocation gave way to crew Titans from Walworth but Holloway continued to run its scheduled RMs and unscheduled RMLs. Route 5 saw the start of an influx of Titans at West Ham which was to persist over the next three months or so, but Upton Park continued to provide Routemasters and was scheduled to lose its share of the route with one-manning later in the year.

An innovative feature on 18th May was a revamping of route 15 (renumbered 15A) and 23 (renumbered 15) along with an in-depth publicity drive to promote the route and its links to the West End, City and Tower of London. One feature of the scheme was a distinctive livery which took the form of a yellow band in place of white and also special yellow and black blinds. Most of Upton Park's RMLs fairly quickly gained their yellow paint-work and six (RMLs 893, 2309, 2523, 2527, 2641, 2737) also sported yellow roofs. As a high profile and easy visible means of recognition the yellow roofs were excellent but they were an initiative on the part of Forest district which was quickly quashed and standard red was restored during June. However a more modest yellow roof band was allowed to survive on RMLs 2402, 2738 (and later 2760). Unfortunately the success of the 15's publicity drive led to the adoption of further routes starting later in the year with the 52 to form a tourist route network, but instead of adopting a distinctive colour scheme for each, yellow was universally applied and in due course the scheme lost credibility.

Route 15's emergence as a tourist route was enthusiastically hailed by Upton Park's staff with the application of yellow roofs in addition to the approved yellow band. RML 893 was one of six in this short-lived style. More demure but less attractive yellow roof bands as on RML 2738 were allowed to remain. Mike Harris/ R.J. Waterhouse

Below A combination of yellow relief band, yellow blinds and special front and rear posters was decided upon for services designated as 'tourist' routes. RM 289 is seen at Hyde Park Corner in April 1986 on the first route to receive this treatment. Colin Brown

The second of the year's large opo schemes was implemented on 3rd August. Routemasters were lost on this occasion from routes 48, 71, 74, 77A, 243 and the 172 was withdrawn completely. Since many of the deposed vehicles were RMLs the opportunity was taken to transfer surplus members of the class to Muswell Hill and Finchley for route 43. Stemming from the various other changes and reallocations which took place on this date came a requirement for the new Plumstead garage to work route 53, giving it the honour of becoming the last garage to receive a first-time allocation of Routemasters; quite an achievement 26 years after they went into production! The loss of crew working on route 77A resulted in the hotch-potch of RM, RML, D and V which had worked it transferring across to the 77.

On the 5th October Norbiton's route 65 began a progressive changeover from RM to M whilst, with 2nd November in mind as the date for opo conversion, Wood Green had begun receiving from the end of September new Ms intended for route 41. However local staff preference was to employ them until the conversion date on route 29 whose WN allocation temporarily became all-M before its sudden reversion to RM on 2nd November. Route 5 was also converted at this time as was Peckham's route 63 and Brixton's 133. An interesting reallocation placed part of route 36B into Catford, taking some of the special BUSCO RMs with it, and contradicting one of the main reasons for selecting the 36 group for

BUSCO, namely that it was all worked from one garage. Walworth, Poplar and Battersea garages closed, and there was sporadic industrial action beforehand, especially from staff at the last-named, illustrating the depth of disillusionment and despair for the future that many felt. The loss to other operators of half the routes in the first batch of LRT tendering (the first four of which changed hands on 13th July) hardly helped.

Sidcup's last bastion of RMs, route 21, had for some time been a semi-regular haunt for Titans, but from 13th November new Metrobuses began flowing in to prepare the route for subsequent opo, the Ms being a temporary expedient at what was intended eventually to be an all-Titan garage. New Cross' RMs (and unofficial RMLs) began replacement by Titans on 26th October but neither garage was quite complete by the end of the year. On Christmas Day a severe and unexplained fire ravaged Southall garage – one of those due for closure – destroying and seriously damaging several Metrobuses. In order to provide replacement rolling stock ten RMs whose London careers were considered to have ended were hastily taken from sales stock and, after receiving attention at workshops in the closed Turnham Green garage, were despatched to Norbiton as temporary stand-ins on route 65 until its crew operation ceased on 1st February 1986.

The first of what was to become a host of Routemaster re-registering (much of it by subsequent owners as well as by London Buses) occurred in September 1985 when RM 456 was

re-registered after withdrawal from service as XMC223A. By prior arrangement, the original number was transferred to a private car belonging to the operating manager of Ash Grove garage who held an affection for RM 456 as the first Routemaster he drove in service replacing trolleybuses. A month later RM 100 became ALA814A to enable VLT100 to go to Don Allmey's AEC Reliance coach. Neither of these two ran for London Buses with their new registrations although later examples did. Autumn saw the start of a huge programme for the fitment of advertisement frames to most double deckers in the fleet. Similar to those tried out earlier at Shepherds Bush, they were scheduled to be placed on all Routemasters still on bus service except those with illuminated offside advertisement panels (very few, if any, of which were still used as intended). A group of London Buses' own inside staff tendered for the work which had a deadline for completion at the end of 1985 which they singularly failed to meet.

1986 marked the 30th anniversary of entry into service of RM 1 and a special exhibition commemorating the event was staged at the London Transport Museum in Covent Garden from 21st May through to January 1987 with RM 1 present. However in a year when so much was happening it remained fairly low key, being overshadowed by other, more momentous events. This was the year when the minibus came on the scene in a big way, not just in the suburbs but also in the heart of the West End. It was also deregulation year

In August 1986 Chalk Farm's Routemasters drifted away as newer vehicles were received for routes 24 and 68, adding another to the growing list of garages stocked entirely with rear engined buses. RML 2373 purrs through Camden Town on 28th August.
R.J. Waterhouse

for the bus industry outside London and the abolition on 26th October after more than fifty years of road service licensing filled the future with uncertainty. However it also opened up fresh fields, even for London Buses should it be enterprising enough to grab the opportunity, although only one RM route penetrated beyond the London boundary into the free-for-all zone; this was the 279 with its northern terminus in the Hertfordshire town of Waltham Cross. A litany of contraction on London Buses, including the loss of many more routes through tendering, resulted in the closure of no fewer than five garages (Edmonton, Loughton, Southall, Bexleyheath and Elmers End), although a portent, perhaps, of things to come was the opening of the new Roundabout base near St Mary Cray and of Westlink on the north Feltham trading estate. The abolition of the GLC on 31st March broke a link with London Transport's recent and hectic past.

Edmonton garage closed on 1st February. With its capacity much under-utilised since trolleybus days because of a restriction on the size of its diesel bus fleet for environmental reasons, its closure at the completion of enlargement at nearby Enfield meant the loss after very many years of the famous destination EDMONTON TRAMWAY AVENUE. (However the tram connection was not entirely lost to London as RMs in the west still displayed ACTON TRAM DEPOT as a reminder of an activity which ceased there half a century ago. Driver-only conversions on this date included the 65 mentioned earlier, as well as

Stockwell's route 77 and the Tottenham allocation on route 171 which became 171A when Metrobuses took over from RMs. A progressive conversion to RML on route 73 started in February with the Shepherds Bush allocation, and though the larger vehicles were specifically required to meet inadequacies on the 73 they also appeared on the 12 and occasionally on the 49. From the end of April, Clapham's RMLs on route 37 began switching to the 137 when Ms arrived as a prelude to one-manning on 21st June, and Hounslow's began departing from 30th May. Other routes being prepared for their new fate were the 155 for which Merton began receiving Ds from early May and Muswell Hill's 134 which was turned over to Ms between 10th and 19th May.

The second main opo conversion programme of the year was on 21st June 1986 when route 2 was reduced in length and converted from RM at Stockwell to M at Norwood. Bow lost a few RMLs when the 8A was taken over by opo Ts whilst route 35 gained Ts at Camberwell and a mixture of Ms and Ts at Ash Grove in place of Routemasters. Route 37, which was by now all-M from Hounslow and Clapham, retained its RMLs from New Cross right up to conversion day when this garage was scheduled to lose its share of the service. Surplus RMLs were used to re-establish the class at Stamford Hill on a partial allocation alongside standard RMs on route 253. Major cross-London services were no longer sacrosanct from one-manning despite the fears of earlier years, such was the pressure of the financial squeeze

imposed by LRT on London Buses, and this was witnessed by the preparative introduction of Ms at Stamford Brook on route 27 from 2nd July, at Croydon on route 68 from early August and at Victoria on the 52A later in the month. Titans replaced RMs at New Cross on the 171 and a mixture of Ms and Ts began taking over at Chalk Farm from RMLs on the 24 and RMs on 68.

A third big one-manning scheme came on 25th October coinciding with the closure of Elmers End garage which retained RMs to the end for route 12 on which it had been a major force for very many years. Routes converted to opo which retained their RMs up to conversion date were 52 at Willesden, the Holloway portion of route 27 and Hendon's route 113, although – perhaps largely through local enthusiast initiatives – routes such as 24, 47, 52A, 68, 155 and 171 also saw Routemaster workings in the days before conversion. The availability of further RMLs allowed the class to return to Tottenham to begin taking over route 73. The year ended with Hounslow's 237 and Enfield's 149 (and unofficially 279) converting from M from November and December respectively. RM withdrawals continued unabated, and once again in excess of 300 were disposed of during the year. Such had the rate of casualties been in recent times that only about thirty Leyland-engined examples still survived in service.

1986 saw a large expansion of 'yellow band' RMs as part of the so-called Tourist Bus network which involved Routemasters on routes 1, 6, 8, 11, 19, 25, 29, 38, 53 and 55. However some routes were never completed (on route 29, for instance, Palmers Green and Wood Green adopted the tourist network features, but Holloway could not do so because its vehicles worked too many other services), while instances of vehicles bearing stickers for one route and appearing on another were commonplace. In the end the problem caused by route-binding buses, plus general apathy, resulted in the downfall of the scheme. The need to economise by decentralising brought the end of two much older London Bus traditions, both on 14th November. One was the closure of the world-respected Chiswick training school and its famous skid patch which had been in existence since January 1925; the other was the end of the undertaking's central licensing department which had controlled the movement of buses since even earlier in LGOC days. Each of the six districts was now left to do its own licensing, and with no central control the old-established body numbering system and the practice of categorising vehicle sub-types (such as had produced the code 5RM5 and all its derivatives) largely fell into disuse.

Inevitably, garages operating on 'tourist' routes had difficulty at times in keeping the buses with route publicity on the right route. RM 241, newly fitted with an advertising frame which in this case surrounds a paper advert stuck to the bodywork, is seen in Willesden. Capital Transport

Points of interest during 1987 included the first example of an RM which had been re-registered actually running in passenger service for London Buses; this was RM 1002 at Peckham whose registration had been transferred to Olympian L 261 in favour of OYM368A. RM 1125 featured in trials with an experimental air filter which was fairly soon discontinued as the pipework involved proved a hindrance to Streatham's maintenance staff. Early in the year, in an endeavour to improve the appearance of its opo fleet for staff morale purposes and better public acceptance, Leaside district introduced a revised livery of red, white and black which duly spread to more than 300 vehicles in its fleet. Though not intended for Routemasters, Stamford Hill garage applied the broad central white band and black skirt to RML 2533, but the broad white band did not suit the lines of the vehicle and was quickly restored to normal although the black skirt remained and was later also applied to RM 339 at Palmers Green. RMs 1988, 2076 (the latter a Leyland-engined vehicle) were repainted at Holloway garage on April 1987 with old-style cream central band. From mid-July a new London Buses roundel was introduced in red (outlined in white so that it showed up against the red background) with a yellow crossbar and this quickly became commonplace as did a range of new district symbols introduced in October. In December, following dialogue over the Leaside livery, a new corporate scheme was announced which was mandatory for all except locally identified networks and subsidiary companies. Ornamenting the basic red was a grey skirt and a 2-inch wide central band applied with tape. On Routemasters the wider central band was allowed to remain in deference to the mouldings which already existed, and the grey was confined to the life-guard rails suspended below the bodywork which soon became discoloured to black by road dirt. Though slow to take on, the new corporate livery was well in evidence by mid-1988.

Driver only conversions continued during 1987 but the emphasis now was heavily on elimination of Sunday crew-work to the extent that, by the end of the year, only routes 8, 11 and 12 were scheduled for Routemasters on

Top **Re-registering of Routemasters with non-matching numbers, a favourite ploy with some provincial operators, first became apparent in London when Peckham's RM 1002 appeared in service as OYM 368A. A BUSCO-equipped bus as denoted by lowered offside running plates, it is seen at Grove Park in May 1987.** Brian Speller

Right **Capital Radio may not be everyone's favourite listening choice but passengers travelling upstairs on Putney's RML 890 in November 1986 could hardly fail to notice it. External advertising, and even the front 'via' blind, was all devoted to Capital. Statements such that it enhanced the quality of service to the travelling public could be believed or not according to personal preference, but the system was later extended to a few vehicles of other classes. The radio could be switched off if passengers complained.** J.H. Blake

Route 109, once one of London's most frequent bus services, finally lost its RMs when one-manning occurred in February 1987, but rear engined double deckers in the form of Fleetlines and Metrobuses had already been present for more than two years. RM 2163 is seen on the Embankment. P.J. Relf

Sundays. LRT's annual business plan forecast many more job losses and at the end of February it was announced that four more garages were to close: Hendon, Wandsworth, Clapton and Seven Kings, although the last-named was subsequently reprieved. Clapham ceased its second spell as a bus garage on 7th February when the newly-built Streatham re-opened, its RMs passing to Streatham and RMLs to Brixton. The latter had already received a few RMLs in January which, pending receipt of route 137 from Clapham, had been used temporarily on routes 109 and 133. Re-allocations arising from the Streatham reopening resulted in Merton losing its last Routemasters along with its portion of route 49. With the exception of Hounslow's 237 – on which a few RMs appeared right up to the end – all one-man conversions on this date were of major services into or through the centre namely routes 30 (RMs ex Clapton and Putney), 109 (mainly RMs from Brixton and Thornton Heath) and 149 (RMs and Ms from Enfield and Stamford Hill). In addition the eastern, Holloway end of route 14 lost its

RMLs and became opo route 14A, leaving operation of truncated route 14 solely in the hands of Putney's RMLs. Holloway should have lost all of its RMLs as a result of these changes but retained a few which made regular appearances on routes 19 and 29. Mid-March saw a progressive changeover to Ms on the busy trunk route 207 but a quantity of RMLs remained at both Uxbridge and Hanwell right up to opo conversion day on the 28th. The displaced RMLs found new homes at Norwood on route 2B and at Peckham and Camberwell on the 12. On 6th June route 1 was shortened and converted from RM to opo Ts and Ls and Leyton's route 55, latterly all RML, also lost its conductors, with Ts releasing the long Routemasters to increase the number of these vehicles on route 2B. The closure of Hendon on this date did not

affect the RM family, but Wandsworth's demise on 11th July meant a reallocation of its RMs. The same date saw the loss by Clapton of its last RMs when route 22 was transferred to Ash Grove ahead of Clapton's scheduled closure date of 15th August. Muswell Hill lost its final Routemasters when opo Ms replaced RMLs on route 43, whilst route 49 also ceased crew operation with Ms taking over at Shepherds Bush and Ls and Ms at Streatham. Clapton's closure on 15th August resulted in the usual round of reallocations, notable on this occasion being the withdrawal of Ash Grove from route 11 which thereby lost its RMLs. On the same weekend the number of bus districts was reduced to five when Abbey District ceased to exist, a change which reflected the diminishing size of the fleet.

Wandsworth garage closed as a normal operational base on 11th July 1987 but the plan was already in mind to use it as a replacement for Battersea as permanent home of the London Coaches fleet. Lined up on the forecourt on 9th May, some of them looking less than immediately usable, are RMCs 1461 and 1462, RM 1613, M 969 and RM 273. Joel Kosminsky

26th September saw Enfield's route 279 falling as the next victim to opo, its workings having been a mixture of Ms and RMs at the end. Next came Cricklewood's 16/A on 21st November along with the 253 worked by Stamford Hill and Ash Grove. A progressive input of Ms had occurred on the 16 group since mid-October but the 253 was heavily dependent on Harrow Weald receiving second-hand Volvos and hired Fleetlines under LRT's latest tendering scheme in order to make Ms available for Stamford Hill. The one-manning of such major services as these was now evoking a good deal of public outrage and indignation such as on route 253 whose conversion the local press heralded with the banner 'GRIM FAREWELL' while three local MPs attended a mock funeral complete with coffin. The final farewell was, in fact, mixed with a touch of carnival as the last buses, Stamford Hill Showbus RM 83 duplicated by recently repainted RM 715, were accompanied by a convoy of preserved buses. On the same date an unusual event occurred in the form of a new Routemaster service, Holloway-worked 135 which was in effect the lopped-off northern end of the 137 with a mixture of RMs and RMLs. The spread of RMLs, and consequent decline of standard RMs, continued with the arrival of the longer vehicles at Putney for route 22, Stamford Brook for route 9 and Westbourne Park for route 7, Westbourne Park's RMLs also managing to stage sporadic appearances on routes 28 and 31. There was, indeed, now little stability as to which type of Routemaster appeared where on the 27 remaining routes. It was commonplace for RMLs to appear on RM services whilst supposedly RML only garages often ran the odd RM or two to help cover shortages. The most unusual occurrence of the year came in mid-November when Forest District, who had enterprisingly converted RMC 1515 into an open topper, commenced running it until shortly before Christmas on route 15 short workings between Aldwych and Marble Arch for the benefit of sightseers wishing to gain an unimpeded top deck view of the West End lights.

Top **One of London's most heavily used routes, the 253, went to driver-only operation on 21st November 1987 proving that, unless considerations other than cost-related ones were to prevail, conversion of the whole fleet was now feasible. Huge public opposition to the change supported by three MPs and various campaigning bodies failed to halt it. On the last day of RM operation, campaigners carried a symbolic coffin through Hackney to the now-disused Clapton garage. Stamford Hill's RM 1102 is about to overtake an unscheduled M 1123 in April 1987.** Colin Fradd

Centre **On the only route still scheduled to be worked by Routemasters outside the London regulated zone, RM 1697 stands in Waltham Cross bus station ready to begin its long run to Smithfield. Also on hand is Potters Bar's LS 112 on a Lee Valley Regional Park contract service operated entirely within Hertfordshire. Crew operation was abandoned on route 279 on 26th September 1987.** Ken Blacker

Right **Heading westwards along the Strand, Stamford Brook's RM 6 is seen shortly before route 9 finally succumbed to the higher capacity RML class.** P.J. Relf

New Routemaster worked services have been a comparative rarity in recent times and the first to be introduced under London Buses' auspices was the 135 on 21st November 1987. For the first time in many years upper case lettering was used on Routemaster intermediate point blinds as demonstrated by Holloway's RML 2620 in these two views at Marble Arch. The last such blinds from the pre-1961 era had ceased to be used in 1976 at Stamford Hill. The route numbers on the side and rear blinds are an inch deeper than the eight-inch high figures normally used. The rear view also shows the size adopted for the rear advertisement frames. R.J. Waterhouse

Left **By 1987 the appearance of many Routemasters left much to be desired, extended periods between repaints having taken their toll. The temporary absence of an offside advertisement on RML 894 reveals the damage often caused upon removal of vinyl advertising posters prior to the fitting of frames.** Stephen Madden

Right **The Company's Wandle district was responsible for a number of unusual one-off Routemaster operations from 1985 onwards, principally for enthusiasts. A noteworthy feature was that correct blind displays were always provided. On Sunday 8th May 1988 RML 2615 worked in place of an omo bus on Surrey County Council service 522 on which it makes a photo stop at Reigate. A Sutton garage plate was carried even though Sutton never had RMLs of its own.** Colin Fradd

The external condition of much of the fleet, but particularly Routemasters, was now very shabby; indeed some RMLs had not been repainted for up to seven years and were a disgrace. The closing of the Aldenham paint shop and of the centrally-managed system for ensuring a rolling programme of repainting had contributed to the decline, which had been further speeded by repeated financial squeezes on the district administrations which could only be met by such devices as abandoning painting programmes. From late in 1987 and through the first few months of 1988 a determined effort was made to redress the situation with a Routemaster refurbishment programme embracing some 139 vehicles, mainly RMLs. Vehicles were despatched to a variety of contractors for repainting, namely BEL, Gatwick Engineering, Kent Engineering, Locomotors, MCW and Southdown Engineering. It was now acknowledged that a core of Routemaster services would survive into the forthcoming but as yet undated era of London bus deregulation, and it was prudent to ensure an adequate number of serviceable vehicles for this purpose. Back in the latter part of 1987 the number of RMs in use actually increased at one stage when 35 vehicles were removed from the Sales department's stock for use as training buses, principally to replace elderly DMS trainers which had to be refurbished and pressed into passenger service to provide rolling stock for tendered operations at Norbiton.

By 1988 it became common knowledge that the existing five-district setup was to cease and that a number of smaller organisations would be created with company status in readiness for deregulation and the privatisation which was widely expected would quickly fol-

low. In February it was announced that there would, in fact, be eleven of these subsidiary companies – to be known initially as units – but excluding Westlink (the tendering subsidiary Stanwell Buses Ltd) which would not be absorbed as would the other tendered units, and the Tours & Charter fleet which would be formed into a separate company. The new companies, it was said, would be encouraged to compete against each other whilst observing maintenance of the London bus network as required by LRT. It was against this background that Bexleybus began operating on 16th January 1988 and Sidcup garage closed. The same date saw the conversion to opo of another two major services, route 25 worked by Bow and route 53 shared by New Cross and Plumstead.

The time had come for a major rethink over the future of the Routemaster fleet which now numbered below one thousand required for passenger service. Few of the 25 remaining Routemaster services were considered suitable for one-manning, except perhaps on their outer fringes, and with a total embargo then prevailing on the purchase of new double deckers ahead of deregulation, no new rolling stock was likely to be received to replace them anyway. Furthermore, deregulation had produced pockets of reversion back to crew operation in the provinces – usually employing Routemasters made surplus from London Buses – and who could tell whether this might not also happen in London after deregulation day? True the newest Routemaster was now twenty years old, but examination of the main structures and particularly the bulkheads, which might have revealed a particular area of weakness, showed them to be sound. Body interiors needed modernising, particularly to

the extent of fitting up-to-date lighting, and rewiring might be desirable. Less satisfactory was the state of the steel structures supporting the running gear, particularly the B frame, where welding of cracks could not go on indefinitely but prohibitive tooling and production costs would rule out the manufacture of the new B frames likely to be required from about 1993 onwards. Another problem for the future would lie in such items as the gearbox cases and engine crankcases where serious supply problems could be experienced through requirements being too low to make it worthwhile to manufacture castings. A much-debated idea was to find a modern replacement for the venerable AEC 9.6 engine which, though excellent, was subject to becoming worn out just like any other.

Earlier, LRT Bus Engineering Ltd (BEL), the Chiswick offshoot of London Transport's own former engineering department, had acquired RM 2175. The company had developed contacts with DAF in Holland as a result of which it was decided to fit one of this company's engines into a Routemaster as a joint venture. RM 2175 was duly converted with such a unit but the matter was not pursued vigorously. Things moved much faster in 1988 when London Buses actually equipped two RMs with new types of engine and placed them into trial service. DAF figured in one of the trials in which RM 545 – a vehicle from the sightseeing tour fleet – received an 11.6 litre naturally aspirated DK1160VS engine rated at 145bhp as against the 115bhp of the old AEC unit. A redesigned sump and flywheel were required and it also proved necessary to feature a revised exhaust manifold, header tank and engine mountings. The re-engining of RM 545 was supervised by Wandle District under whose initiative a Fiat-built 8.1 litre Iveco type 8361 engine was fitted into RM 1894. This engine was a naturally aspirated version of the marine diesel with which 100 of the District's B20 Fleetlines had been successfully re-engined, and it had the advantage of being less expensive than the DAF. The conversion work on RM 1894 was carried out by Scottish Bus Group Engineering Ltd who were located conveniently close to Glasgow-based Industrial & Machine Diesels, the UK concessionaire for Iveco's industrial division. Both buses duly entered service on route 2B from Norwood garage although in January 1989 RM 1894 transferred to Victoria.

Two attempts were made during the year to modernise interior lighting. One involved RM 470 which was fitted by Locomotors Ltd at Andover with fluorescent tubes on both decks. Upstairs the tubes were plain but on the lower saloon they were concealed by *Transmatic* advertising panels in a neat arrangement. The *Transmatic* panels were self contained modules, each complete with inverter, tube and local wiring, enabling the system to be installed without major changes to the vehicle's original wiring. At the same time a revised saloon heating system was installed on the lower saloon with below-seat heaters fed from the vehicle's water system. A much-needed new design of wiper motor was also installed. RM 470 entered service in its improved guise at Victoria on 4th August. In a totally unconnected experiment Peckham's RML 2613 was equipped with *Transmatic* panels concealing new fluorescent tubes on both decks. However in this case, unlike RM 470 and a group of ten Cardinal Metrobuses dealt with earlier, the panels were less neatly installed, indeed on the upper deck the offside ones did not even align with those on the nearside. Minor modifications during the year included the fitting on a number of RMLs of Fleetline-type front trafficator panels instead of the Routemaster 'ears'. RML 2275 at Camberwell became the first of its class to receive non-opening front windows after an accident, using parts from a withdrawn RM which was being cannibalised at Camberwell. A move in the other direction occurred at Holloway when derelict RM 57 surrendered its opening front windows to make good accident damage on early-bodied RM 98 after it came into contact with a crane on a low loader! This was the eighteenth early RM body to be dealt with in this way.

During the year Holloway garage continued to repaint Routemasters with cream bands, including RMs 804 and 2009, RMLs 885 and 2413 and trainer RMCs 1476 and 1494. Perhaps in recollection of an earlier experiment at the garage, it even turned RM 1278 out with a red band. At Camberwell RM 1666 appeared with a full grey skirt but was the only one so dealt with. On 21st January RM 1681 was leased to LBL subsidiary Stanwell Buses Ltd, primarily for use as a trainer, and in due course appeared sporting the Westlink fleetname in place of the London Buses roundel. After a short lull following the end of the 1987/8 intensive repainting programme, another started and by the end of the year a high proportion of the Routemaster fleet was looking smarter than it had been for a very long time.

As it cruises through Victoria, hints of the major change which lies below the bonnet can be perceived as the sun shines through the front grille of RM 545 giving a glimpse of modifications necessary to suit its new DAF engine. Stephen Madden

The diagram for route 11 inside Victoria's RM 470 makes use of the bulkhead advertisement panel which for many years has mostly been left underemployed. More noteworthy, however, are the Transmatic light fittings which combine a welcome improvement in lighting level with panels into which card mounted adverts can be inserted. Joel Kosminsky

On 4th May 1988 LRT stated that no further routes were to be converted to opo other than those already announced and that Routemasters would remain for an indefinite period. The wheel of fortune had turned full circle for the Routemaster, but too late to save almost two-thirds of the original fleet which had moved on to pastures new or, in the majority of cases, to the Barnsley scrapyards. At the time, in fact, no further opo conversions had been announced but four more were intended covering routes 28, 29, 31 and 135, but the startling news soon broke that the 28 and 31 were to be converted to a high frequency midibus operation using seventy Alexander-bodied Mercedes 28-seaters. Meanwhile other service changes continued to take place including the introduction on 13th August of new, RM/RML-worked route 10. This was in effect the western arm of the old established 73 which was diverted to Victoria and in doing so replaced the pioneer Red Arrow service 500. On the same weekend the very last Sunday crew service, route 12, was converted to driver-only operation. The only Routemasters now working on Sundays were sightseeing vehicles.

Hornchurch garage closed on 24th September as a result of extensive tendering losses, but in remembrance of better times RM 1676 and RML 2288 were both borrowed on the last day to work on route 165. Though occasional outings by crew buses on opo routes were perhaps no longer the novelty that they used to be, one case which hit the news occurred on 18th June when RML 900 worked from Finchley on route 26. It was newsworthy because RML 900 was now owned by Clydeside Scottish Omnibuses, having been sold by London Buses as being beyond economical repair. On 10th September Finchley's RML 903 – one of the few surviving Routemaster Showbuses – repaid the compliment by working on Clydeside's route 23. A particular blow to morale was the loss of route 24 to Grey-Green from 5th November, and on the last day of the old regime crew operation returned when RM 804, borrowed from Holloway, worked on the service with Chalk Farm garage's Operating Manager herself working as one of the conductors. Staff availability encouraged Shepherds Bush to return RMLs to routes 88 and 220 on August Bank Holiday

Monday whilst, even more unusually, Holloway revived part of the defunct 104 from 28th to 30th December together with six RMLs per day. The last opo conversion of the year coincided with the loss of route 24 and meant the end of RMs (and in the case of Holloway occasional RMLs) from routes 29 (worked until conversion by Holloway, Palmers Green and Wood Green) and 135 (Holloway alone).

Above **RML 2721 demonstrates the flatter, 1984 style of bonnet grille as it crosses Oxford Circus on route 10, newly introduced on 13th August 1988 and noteworthy, until a subsequent rescheduling, for its unreliability. The destination screen is barely deep enough to accommodate a legible two line display.** Tony Wild

Although not due to begin functioning as separate legal entities until 1st April 1989, the eleven new unit managements took over control towards the end of 1988, units 1 and 2 (London Central and Selkent) on 7th November and the remainder on 5th December. From 3rd December the new names and insignia began to appear at most units, in the case of Routemasters above the lower saloon windows towards the front of the vehicle. All were in white and employed similar italic-style lettering in deference to a management decision that the eleven companies should all adhere to an identical livery style, at least until deregulation, on the grounds that a uniform image represented their strongest selling point. Only at the Centrewest and Metroline units were signs of the new management still not visible on buses at the close of 1988, finally appearing in January and February 1989 respectively.

The early months of 1989 were the quietest for many years on the Routemaster front and it was not until 4th March that anything of note happened. This was the day when Westbourne Park's Routemasters on route 28 made way for the Gold Arrow fleet of MA-class Mercedes midibuses in the most revolutionary opo conversion ever to take place in London. On the same date alterations took place on route 15 of which Westbourne Park now gained a share to partly counteract its large loss of crew work, and at the same time a new 15B commenced with RMLs from Upton Park. Being a Monday to Friday operation, the 15B's first day was 6th March which was also noteworthy as being the starting date for another variant on the 15 theme, the X15 'Beckton Express' with RMCs (see chapter 11).

Above **Few Leyland engined RMs survived through to 1989 but Victoria's RM 2015 was one of those that managed to do so. The unofficial Leyland badge on the full depth front grille is a sign that local initiative still survives. The old illuminated advert panel now serves as a frame for vinyl ads but is no longer lit up.** Tony Wilson

Below **London Central's Camberwell based RML 2275 (incorrectly carrying the classification RM) typifies the results of a large scale 1988 repainting programme which successfully improved the external appearance of many vehicles in the fleet but without reaching the high standards of paint finish once obtained from Aldenham overhauls. The plain upper deck front windows normally associated with early RM bodies were installed after an accident and are unique to this RML.** Tony Wild

On Saturday 1st April 1989 London Buses ceased to operate buses directly and day-to-day operations were henceforth controlled by the new subsidiary companies which had been registered in January. The names and addresses of the new companies were promptly applied to the legal ownership panel on all vehicles but the familiar London Buses roundel was also kept and gave continuity with the past. All major assets, vehicles and garages remained vested with London Buses Ltd to whom the subsidiaries in effect paid rentals for their use.

At the coming into existence of the new companies the future for London's surviving Routemaster fleet looked likely to be an exciting one with no sign of any immediate diminution in their role. With the likelihood of an expansion of the trials with new engines and further major attempts at body refurbishments, the determination seemed to exist that Routemasters could be expected to play a significant role well into the 1990s and perhaps even into the 21st Century.

From the immaculate to the down at heel; both extremes are to be found in the 1989 Routemaster fleet as it heads into the new decade. Palmers Green's RM 5, retained primarily for special duties, glistens as it approaches Waltham Cross on loan to Potters Bar for a special Routemaster day on the commercially registered 310 group on 23rd September 1989. Leyton's RML 2377 looks anything but well cared for in July 1989 with dented dome and bonnet top, overpainted radiator grille with missing badge, and lopsided registration plate. At Lewisham bus station Catford's RM 400, minus its fog lamp, shares the stand with Titan T 983, Mercedes StarRider SR 70 and a Leyland National. This mixture symbolises the position at which, through force of circumstances, the London Buses fleet has now arrived. Vehicles of vastly differing types, covering several generations, run side by side in a welter of non-standardisation which is the very antithesis of what the Routemaster stood for. R.J. Waterhouse/Joel Kosminsky/R.J. Waterhouse

The 11 new wholly-owned companies were granted semi-independence in certain spheres though not in the use of liveries, but this did not deter two of them from proclaiming their existence with specially painted vehicles. Perhaps appropriately, the two companies concerned were those which had chosen to resurrect for themselves titles from the past, London General (full title London General Transport Services Ltd) and London United (London United Busways Ltd). London General scored a real triumph when, on its first weekday of operation, Monday 3rd April 1989, two RMs spendidly adorned in General livery attended a special launch at Victoria bus station along with preserved S 742, after which they set off for suitable refreshments at the London General public house in Moorgate. RMs 89 and 1590 were in basically the same red and white livery with silver roof and black lining as the four 1983 vehicles although with minor differences, notably the omission of a thick black band above the lower saloon windows. Allocated to Victoria garage, they quickly became a familiar sight in the West End and elsewhere on this garage's several Routemaster routes. London United's first revival employed Metrobus M 1069 which was repainted in what purported to be a version of the earlier company's tram livery. In May 1989 it was joined by RML 880 whose somewhat different version of the red and cream livery aspired to copy London's first trolleybuses, the famous 'Diddlers' of 1931. Red lower and between-deck panels, and red lower saloon window frames, were relieved by cream bands above and below the downstairs windows and around the upper deck ones. Black edging, plus silver roof and wheels, completed the ensemble. The result was cheerful, if perhaps historically less authentic than the London General pair, but the bus was slow to enter service in its new guise. North Weald rally day of 18th June 1989 saw it at work on special service 339 but without fleetnames or numbers; regular service use from Shepherds Bush on routes 9, 10 and 12 did not commence until mid-summer, by which time the vehicle carried the fleet number ER 880, recalling the identity under which it had been delivered in 1961 prior to the adoption of RML as the type classification for 30ft Routemasters.

Above **Though their GENERAL fleet names were perhaps a little too large, the two Victoria based Routemasters made an excellent advertisement in 1989 for the new London General company. By a quirk of good fortune for the photographer, both are seen heading into Victoria bus station in May with RM 89 in the lead. RM 89 is one of a small number of RMs which, over the years, have acquired front wheel hubs with embossed AEC motif, originally on early post war RTs and much older than the Routemasters themselves.**
G.A. Rixon

By adopting the old London United name the new 1989 company established a long severed link with the past which was further enhanced by the revival of the old company's crest. This is seen adorning ER 880 at Hyde Park Corner whilst allocated to Shepherds Bush. Russell Upcraft

South London celebrated its first day of existence on 1st April 1989 by projecting a route 159 journey southwards to Godstone. The vehicle selected recalled the use of red RMLs, including RML 2305, at Godstone when new nearly a quarter of a century earlier. With part of its offside advertisement display missing and a deep dent in the nearside corner of the front dome, RML 2305 was perhaps not the tidiest choice with which to usher in a new era. Russell Upcraft

CHAPTER FOURTEEN

LONDON SIGHTSEEING

Back in 1951, the year of the still-remembered Festival of Britain, London Transport inaugurated its famous Round London Sightseeing Tour using four RTs which had travelled across Europe a year earlier and which still proudly carried their GB plates and commemorative plaques. The service was a huge success and was perpetuated annually thereafter, eventually becoming a year-round operation as tourism grew. A quarter of a century later other operators, scornful of London Transport's complacent and lacklustre approach, were taking 60% of the trade and threatening to capture even more unless something was done to reverse the trend.

In 1986 new life was breathed into the operation. In an imaginative move, newish Metrobuses were swept off the tour by a fleet of 50 Routemasters, marketed under the title of Original London Transport Sightseeing Tour and painted in traditional Routemaster livery with cream relief and gold lettering. Furthermore, twenty of the RMs had open tops. This was a shrewd move against the opposition whose widespread use of open-toppers was one of the main reasons why they had gained so much ground. Conversely London Transport's own flirtation with a small batch of ex-Bournemouth DMOs and a miscellany of vehicles hired over the years had always appeared less than half-hearted. The former Tours &

Charter department was revamped in January 1986 as the Commercial Operations Unit, a separate profit centre within London Buses Ltd, running from its own premises and responsible for its own vehicle maintenance.

Prior to 1986 the only recent regular Routemaster involvement on the London sightseeing tour had been with rear-engined FRM 1 which had demonstrated its superiority over the Fleetlines between February 1978 and February 1983. In addition ex-Northern General Routemasters, in closed and open top guise, had sometimes been hired through the late Prince Marshall's Obsolete Fleet. The fifty 'new' tour Routemasters comprised 11 RCLs, which were all that remained in stock and were ideal for their new task because of their comfort, 19 closed-top RMs and 20 open-toppers:-

RCL	2220, 2235, 2240, 2241, 2243, 2245, 2248, 2250, 2258, 2259, 2260
Closed-top RM	163, 237, 307, 377, 450, 460, 479, 545, 572, 597, 644, 710, 753, 778, 785, 811, 850, 1919, 2050
Open-top RM	49, 68, 80, 84, 90, 94, 143, 235, 242, 281, 313, 398, 428, 438, 562, 658, 704, 752, 762, 925

Above The first Routemasters to see passenger service with operators who had not run the type since new were former Northern General vehicles which, ironically, came to London and could be found on the sightseeing tour from 1978 onwards in full London livery. Sometimes the garage code AH was displayed, this being the one-time London Transport Nunhead premises at which Obsolete Fleet Ltd was based, who were contracted to supply buses 'on hire' to supplement London Transport's own tour DMSs. With an Obsolete Fleet ex-Midland Red BMMO D9 open topper about to pass, Brakell-owned ex-Northern 3102 stands at Victoria carrying full London indicator displays front and rear and fleet number RMF 2794 (ex-Northern 3103, 3091 and 3090 being RMF 2791-3 respectively). In the late 1980s this bus was still on London tour work, but as an open topper in Blue Triangle colours. Tony Wild

Top Right Former Northern General 3090, latterly known as RMT 2793, had run in London for a number of seasons prior to coming into London Buses' ownership. It is seen at Victoria after becoming a full time member of the London fleet. R.J. Waterhouse

Centre Right The first of the open-top conversions of 1986, RM 80, is seen just after completion in Aldenham works. Capital Transport

Right On an ideal sightseeing day in June 1987, RM 94 tours along the Victoria Embankment with a full complement of passengers, not only upstairs in the sunshine but also in the lower saloon. R.J. Waterhouse

A contract was placed with Aldenham Works to prepare the fifty vehicles for service, a £250,000 task which Aldenham was pleased to receive having recently lost London Buses' overhauling work. This included a full overhaul and repaint for all vehicles plus the open-top conversions of which RM 80 was the first to be completed. This conversion was very neatly designed, retaining the front window frames and the first two side bays with the second sloping down to waistrail level following the same contour as the rearmost side window had formerly done. At the rear the former structure was retained to just above the emergency exit. Waterproof floors and seat coverings were fitted on the upper deck. All fifty vehicles were fitted with public address equipment for use of the guides which they were to carry, and with a sliding intercommunicating window behind the driver. In the case of RM 2050 an AEC engine was fitted at the old AEC Southall works to replace the Leyland one. The RCLs initially retained their open rear platforms from route 149 days, and from the front the open-toppers were converted to resemble the RCLs with the same two-piece indicator layout. The RMs were converted from full to semi-automatic, bringing them into line with the RCLs and opening up the prospect of savings through improved fuel consumption and reduced transmission wear.

Sporting what appeared to be a particularly good paint job, especially when compared with some of Aldenham's recent output, the first Routemasters entered service on the OLTST on 22nd March 1986, just before Easter. This Routemaster revival approximated with the thirtieth anniversary of the entry into service of RM 1 and was in no small measure responsible for a 5% increase in market share in what was generally a dismal summer for tourism. Despite a preponderance of gloomy weather, the open-toppers proved particularly popular, and it was clear that a larger fleet of such vehicles would have been desirable, so RMs 163 and 1919 were decapitated in time

for the main summer season. To further augment the 22 already in stock, a privately-owned open-top Routemaster was hired from 1st July 1986 onwards. Belonging to J. Leaver of Sutton, EUP406B was one of the Northern General 'RMFs' which had been hired in earlier years through Obsolete Fleet. Carrying London Livery and the fleet number 'RMT 2793', it operated from Battersea for the remainder of the summer season. As part of its expansion programme, the Commercial Operations Unit won a Kent County Council contract for a late evening journey to Tunbridge Wells and back on Green Line route 706 formerly run by London Country South East. On the first evening of operation, 27th October, the regular coach Olympian LC 2 was duplicated by RCL 2259, recalling this class's halcyon days of Green Line operation two decades earlier. This vehicle also operated the service, though this time alone, on New Year's Eve, whilst the final day of the contract – 29th May 1987 – found RCLs 2241 and 2250 both performing a last commemorative run along with RMA 26, one of six ex-airport coaches now in the sightseeing fleet.

It soon became evident that there were too many standard covered-top RMs in the OLTST fleet. During the summer their passenger appeal proved limited as the majority of tourists preferred open-toppers, even on non-sunny days, whilst in winter their open platforms let in the cold, emphasising the need for doored buses on tour work. Indeed, doors were fitted to the eleven RCLs during the autumn months of 1986 although these were not to the same design as the originals. Safety regulations now required a three inch rubber edge to each door plus provision of a buzzer to warn the driver of any fault or blockage. At the end of 1986 six RMAs no longer required by LRT Bus Engineering Ltd (BEL) as staff buses were acquired by London Bus Sales for transfer to the OLTST fleet. Conversion into service condition and fitment of public address equipment was carried out under contract by BEL at Chiswick starting in December 1986 and included the provision of front indicators which were of the RCL type. The vehicles were upholstered in the correct style of grey moquette acquired from London Underground with whom it was still in use on Metropolitan line A-stock, the RCLs being similarly re-upholstered later in 1987. In the case of the RCLs this meant a return to original condition from the blue and green seats carried latterly. The six RMAs concerned were RMAs 15, 22, 25, 26, 51, 65 of which RMA 25 was the first to be completed. Two of the contract – RMAs 26 and 51 – were reputed to be the last LBL-owned vehicles to be repainted at Aldenham before its demise. RMA 26 was noted as the first into service on 22nd March 1987, all six being operational by July.

Two additional open-top RMs joined the fleet early in 1987. Leyland-engined RMs 1783, 1864 had been sold by LBL in May to Rees Industries on behalf of Transworld Leisure of Liverpool for operation at the garden festival held there during the summer. The terms of sale included conversion at Aldenham to open top in the same manner as the OLTST vehicles except that the original three-piece front indicators were retained. Transworld Leisure subsequently went into liquida-

Above **With their new paint gleaming RMs 377 and 597 make a brave sight at the tour pick-up point in the Haymarket in April 1986. Adoption of the old red and cream livery with traditional gold lettering (albeit applied in modern *Systemtext* rather than transfers) produced a classic image which was further enhanced by the tasteful posters. Mercifully the temptation was resisted to fit advertising frames which were sweeping through the rest of the fleet at the time. RM 377 was later converted to open top and RM 597 was replaced by a re-acquired Routemaster open-topper.**
Joel Kosminsky

Centre **For London Coaches' first evening on Green Line route 706 in October 1986, RCL 2259 accompanies the scheduled vehicle, Olympian LC 2, from Victoria on its journey to Tunbridge Wells. The incorrect route number carried by RCL 2259 is a reminder of the days when Routemasters regularly made this run.**

RCL 2241 rounds Marble Arch and shows its new platform doors, fitted in autumn 1986.
Roger Whitehead

tion and London Bus Sales, being one of its creditors, repossessed the two RMs in December 1986. By January 1987 they were in store at Battersea and were subsequently returned to London colours by Kent Engineering, initially with a white relief though this was later repainted cream. It was not until June that ownership was legally established, enabling the two to be officially incorporated into the sightseeing fleet, releasing closed-top RMs 460, 597 and 785 to sales stock as being of equivalent book value.

After lying disused at Battersea since the close of the 1986 summer season while discussions took place as to its future, approval was given in July 1987 for hired RMT 2793 to be taken into stock. In exchange its owner, Mr Leaver, was handed RM 2050 as part of a final settlement over each side's outstanding claims. RMT 2793 returned to the tour late in the summer season, on 31st August, after being refurbished and repainted at Carlyle Works, an RMC-style blind layout now being fitted at the front. A further Routemaster recruit to the Commercial Unit fleet – although not for the sightseeing but primarily as a training vehicle – was RMC 1491 in July 1987. Repainted at Battersea into what was effectively a red version of its original Green Line livery, including cream window surrounds, red wheels, wings and lifeguards, and a gold between-decks bullseye each side, RMC 1491's arrival meant that every extant class of Routemaster except for RML had at least one representative at the Battersea base.

A surfeit of ordinary RMs led to lending of five (RMs 237, 307, 377, 450, 545) to Southend Transport in November and December 1987. Two others, RM 850 superseded by RM 572, were lent to Blue Bus Services of Eccles for operation in the Greater Manchester area in December 1987 and January 1988. Although, with the three acquisitions of 1987, the open-top fleet now stood at 25, it was decided to boost this figure by a further seven for the main 1988 season. Conversion work was split amongst three companies: Carlyle converted RM 237, Carlton PSV dealt with RMs 307, 450 and Kent Engineering handled RMs 377, 572, 644, 753, the first ones returning in their rebuilt condition on 2nd July. Generally similar to previous conversions, these differed in retaining their three-piece front indicator displays. RMs 307, 450 were unique in having been rebuilt with the carriage of disabled people in mind, a wheelchair lift being incorporated into the second and third nearside bays. The lower saloon, now reduced to only ten seats (making the vehicles 46-seaters) retained its longitudinal wheelarch seats and a seat to the rear of the wheelchair lift; the remaining space was capable of accommodating up to four wheelchairs. After this latest spate of open-top conversions only six closed-top RMs still remained under London Coaches' control (RMs 479, 545, 710, 778, 811, 850) of which RM 545 was involved in the experimental fitment of a DAF engine as recorded in chapter 13.

On 4th April 1988 the fleet was officially transferred from Battersea garage – which had remained LRT rather than LBL property and was a candidate for redevelopment – to the former Wandsworth bus garage. Some delicensed vehicles had actually been stored

in Wandsworth beforehand and the move, when it came, occupied several days. Even after it was complete, shortage of space at Wandsworth meant the continuance of open-air parking at Battersea. On 5th April 1988 London Coaches commenced contract commuter services from the New Ash Green area of Kent to London following the collapse of the previous operator, Bexleyheath Transport, which required a few vehicles to be outstationed at Kentish Bus premises in Dartford. RCL 2250 was kept there for a while as a spare. In the spring of 1989 this outstation moved to the former Horlock's coach base in Northfleet.

On 1st April 1989 the London Coaches fleet was transferred to a new LBL subsidiary company, London Coaches Ltd.

Top **RCL type front indicators were a new feature on the six RMAs introduced to the sightseeing fleet in 1987. Upper deck travel is always more popular with the lower deck with tourists, many of whom are visitors from overseas, and this preference was heightened on these front entrance/staircase vehicles. A lightly loaded RMA 25 is seen on the tour in March 1988.**
John Miller

Above **Wheelchair lifts installed in two RMs in 1988 enabled up to four wheelchair passengers to enjoy the 90-minute trip around the capital with operations initially scheduled for Mondays, Thursdays and Saturdays. RM 307 demonstrates the doorway installed in the second bay.**
G.A. Rixon

CHAPTER FIFTEEN
LIFE AFTER LONDON

In 1982, twenty-three years after its first appearance on London's streets as London's Bus of the Future (prototypes excluded), London Transport commenced the wholesale withdrawal of the RM class. This fact alone was a remarkable tribute to the far-seeing concept and sturdy construction of this remarkable design, for how many other major classes of urban bus anywhere in the world has ever lasted so long intact? Within Britain, both in London and elsewhere, many later bus types had come and gone while the Routemaster still soldiered on. Even in 1982 obsolescence and redundancy were the reasons for withdrawal, not physical deterioration.

Inevitably a few vehicles failed to make it through to 1982 where catastrophe had struck and economic repair could not be achieved, but these were only five in number. An early victim was RM 1768 which, while less than three years old and still carrying its original body, was demolished by a flywheel fire when working light through the West End back to Middle Row garage in July 1966. However, thanks to the 'float' system whereby two additional bodies were constructed, RM 1768 did not need to vanish completely from sight but

duly reappeared with another body. The same applied when Highgate's RM 500 (carrying body B560) became a fire victim on 9th January 1968. Burnt-out bodies B1768 and B560 were officially dismantled at Aldenham in April 1967 and October 1968 respectively. With the float of two now used up, it was unavoidable that any further fatal accident victims would have to be written out of the fleet, and the first to fall into this category was RM 304. Working Turnham Green's route N97 in the early hours of 26th November 1971, this vehicle was struck on the offside by a large milk tanker, injuring eleven people and causing thousands of gallons of milk to pour into the Fulham Palace Road and adjacent streets. The badly buckled body of RM 304, B246 was the only one of the early variety with non-opening front windows to be scrapped early. On the morning of Sunday 2nd April 1972 at about 4am a severe fire enveloped a number of RMs in Peckham garage in what appeared to be an arson attack centred on RM 1659. This bus, together with RMs 1268 and 1447 was a total wreck although a fourth vehicle, RM 1436, was found to be salvageable. An unexplained fire similarly gutted to a shell RM 50

(which carried newer body B1746) at Walworth garage on 12th March 1973. The four fire victims and RM 304 were all officially written-off in February 1974 although some of the remains lay around at Aldenham for some time afterwards and most of the top deck of RM 304 was salvaged for the future repair of other accident victims.

Above **Typifying the many non-psv roles to which sold Routemasters have been put is RM 1417. Purchased in December 1984 by Dolphin International Displays Ltd of St Albans and modified as shown on behalf of Interflora, it is seen at the Uxbridge showground during the Middlesex Show of June 1987, curtained upstairs and suitably bedecked. The nearside lower saloon has been converted to an open-sided show case, the hinged cover of which is in a raised position and partly hidden by the awning.** R.J. Waterhouse

With the exception of prototypes RM 1 and RM 3, which were sold in April 1973 to Lockheed and February 1974 to preservationists respectively (RM 1 was later retrieved and taken back into stock) and the one-off RMF 1254 which passed to Northern General in November 1966, the only Routemaster to be sold by London Transport prior to 1982 was RML 2691. This surprise sale on 21st September 1972 of such a comparatively new vehicle was presumably accompanied by a very strong financial inducement to part with it, although in later years when 30-footers were at a premium the loss of even one member of the class was to be regretted. Allocated to Hanwell from new in September 1967, RML 2691 had not yet received its first overhaul and therefore still bore its original body when withdrawn from passenger service on 1st September 1972. After receiving mechanical attention at Hanwell and Chiswick, it left London to be converted by Lex Tillotson of Burnley into a mobile display unit for the Mary Quant products of Gala Cosmetics (International) Ltd of Surbiton. In a mainly yellow livery it was noted touring the USA and Canada in 1973 and has since appeared spasmodically in Scandinavia and other parts of Europe, returning to Britain every so often. At the time of writing it was still fulfilling this function.

The infamous Barnsley area scrapyards, whose rapacious appetite had swallowed up thousands of RTs and RFs, had their first taste of London Routemasters in January and February 1978 when 21 semi-derelict vehicles were dismantled by Wombwell Diesels. Comprising 17 RMLs, 2 RMCs and 2 RCLs, these were the worst of the first contingent of vehicles purchased back by London Transport from London Country, whose restoration to full running order would not have been financially worthwhile. However, as with the RTs and RFs before them, an arrangement was made with the scrappers whereby many nominated mechanical and body items were returned to London for re-use, so to this extent the

deceased vehicles were not a total loss. In addition to these, as recorded in chapter 2, a couple of London Country RMLs went direct to Wombwell Diesels for scrap in March 1979, while in a similar vein Wombwell dismantled two ex-Northern General RMFs for London Transport in 1978 in addition to many other NGT Routemasters which found their way to the Barnsley scrapyards. In 1981 further former London Country Routemasters were dismantled by Wombwells, mostly without having been used after their return to the London Transport fold, as were a few ex-British Airways RMAs.

With Routemaster withdrawals commencing in earnest in 1982 – more than 150 being withdrawn in August and September alone – London Transport let it be known amongst the scrap trade that large quantities would become available and that quotations were required for dismantling and retrieving certain items from them. As a trial run RMs went, one apiece, to seven breakers in August and September to seek the most favourable terms, these being RM 1175 to PVS, RM 1302 to Booth, RM 1337 TBP Engineering, RM 1390 Way & Williams, RM 1442 Paul Sykes, RM 1443 Bird's and RM 1781 W. North. The fact that none of these were early numbered RMs coincided with an adopted policy of disposing in the first instance of vehicles equipped with Leyland engines and/or Simms electrical units. The age of the body was not considered important since even the earliest ones were still in excellent structural condition. In the early days this policy was vigorously pursued to the extent that Leyland engined RMs comprised some 87% of all those disposed of in the first couple of years. Later it slipped somewhat as other more immediate criteria dictated disposal such as accident damage, defective mechanical or body units, or even an imminent Certificate of Fitness (later Freedom from Defect) expiry where, with pressure on labour or materials, it was more expedient to sell a vehicle rather than recertify.

W. North of Sherburn-in-Elmet, Leeds, won the contract to scrap the first fifty RMs, reviving memories of early post-war days when vast quantities of old London buses had fallen victim to this company's torches and sledgehammers. In addition it was arranged that one hundred vehicles would be dismantled at Aldenham in late 1982/early 1983 by staff of Vic Berry of Leicester, although the number actually dealt with fell short of the full total by a couple. Later scrappers such as Booth of Rotherham dealt with a number of RMs but far and away the greatest handler of these vehicles has been PVS (Barnsley) Ltd of Cudworth who by spring 1989 had disposed of not far short of 800 Routemasters.

Although the scrapyard would inevitably be the fate for the great majority of redundant Routemasters, it was hoped right from the start that some would move on to a second career thus greatly enhancing their sale price. No obvious home market existed for them as

A hinged offside door, additional front lights and other modifications are visible on the former RML 2691 as it receives attention in the Hounslow premises of LPC Coachworks in 1975. Colin Brown

PSVs, except perhaps for contract or factory work, because of the countrywide move away from crew to driver-only buses. However small numbers began to find new homes and, most interestingly, a demand from overseas quickly established itself. Indeed two of the very earliest withdrawals in August 1982 (RMs 326, 496) were set aside at Ensign's Purfleet premises – where many early RM withdrawals were stored – against a contract from Japan. Sometimes the exported vehicles were converted internally for display purposes etc, but in a number of instances, and most notably at Niagara Falls, they continued in passenger service as tour buses emulating the many RTs and provincial Bristol Lodekkas which had gone there before them. Within seven years of the commencement of Routemaster sales examples of these traditional London double deckers had appeared in some 26 countries covering every continent in the world.

Preservationists, both at home and abroad, were quick to seize the opportunity to acquire Routemasters to ensure that, even though numerous examples of the type were to operate in their natural habitat for years to come, good examples were sure to survive for posterity. One of the earliest preservation schemes, dating from January 1983, saw RM 1414 acquired by the Greater Manchester Transport Society for exhibition in the

Museum of Transport in the city where the original vehicle of this number had operated temporarily when new. London Transport itself, whose far-sightedness in the preservation of representative vehicles from the past was legendary, preserved standard RM 1737 and, as a typical representative of a bygone Green Line era, RCL 2249 (a late substitute for RCL 2260 which was found to be in less good condition). These vehicles passed out of the service fleet into London Transport Museum stock in December 1985 and April 1984 respectively, although neither has yet been on public display at Covent Garden.

Having had vast quantities of AEC Swifts and Merlins and Daimler Fleetlines to dispose of in recent years, London Transport set up its own London Bus Sales department who, with large numbers of Routemasters now becoming surplus, decided to convert one as a promotional vehicle. Former Mortlake garage Showbus RM 1563 was repainted in 1985, complete with gold fleet names, cream band and relevant bus sales propaganda, as the sales department's flagship. Little was it realised at the time that, within a couple of years, all serviceable redundant Routemasters would be eagerly sought after by potential buyers. The trigger that was to revive fortunes by giving them a second working life in many cities and towns throughout the United Kingdom was

the Government legislated bus deregulation which took effect from 26th October 1986. For good or bad, deregulation, coupled to the privatisation of the former subsidiaries of the National Bus Company, changed the face of bus operation beyond recognition. Whilst mini and midibuses spread like wildfire across the length and breadth of the nation, generally at the expense of conventional buses, the other end of the equation found an unexpected resurgence of crew operation as a competitive tool, for which traditional open platform double deckers were mostly preferred. As most other open platform buses had by now been consigned to the scrapheap, this left London's famous Routemasters as virtually the only vehicles available for operators wishing to return to the operational methods of days gone by. Despite their advanced age, and the mechanical complexities to which many provincial operators were unaccustomed, there was suddenly no shortage of operators wishing to speculate on the future by introducing Routemasters into their fleets. Soon Routemasters began appearing in many unaccustomed liveries, recalling the days of the RTs when these, too, had found advocates far and wide. In several instances RMs met up with ex-London DMSs many of which had also found a second and sometimes more satisfactory existence away from the capital.

Only a small minority of RMs withdrawn by London Transport prior to 1985 escaped the scrap torch, and few of those which did so were used for further passenger carrying work. An exception was RM 1878 which was sold in August 1984 for use, still in home territory, as a courtesy bus running a free service from Aldwych to the Hyper Hyper fashion store in Kensington High Street. In striking pink livery it was unlikely to be confused with service buses along its line of route. A later, less glamorous job, was as a playbus with the London Borough of Wandsworth. RM 1878, with 'running number' HY1, is seen at the Kensington end of the free service. Colin Brown

RM 2120 takes London's red livery to the Continent in the guise of a mobile theatre box office. Its old roof has been sliced off at the cant rail and a new flat one, complete with translucent panels, has taken its place. This can be lifted for static use in suitable weather; when lowered the bus is just 4 metres high. Registered in West Germany, the bus is seen in Amsterdam in April 1988. Joel Kosminsky

The best views of Niagara Falls are from the Canadian side, and 'de-luxe sightseeing tours' using Routemasters have been a popular feature for a number of years. Double Deck Tours Ltd first purchased Routemasters in 1983 followed by others at approximately two yearly intervals. This member of the fleet still bears London red livery. Tim Clayton

Stagecoach Ltd caused raised eyebrows in January 1985 upon purchasing RMs for operation on its services in and around Perth and Dundee when the overall trend was still firmly against the use of buses requiring conductors. However the Magicbus image did not appear until Deregulation Day in October 1986 which was infamous for bringing certain Glasgow thoroughfares to a standstill through the sheer number of buses attempting to use them. Stagecoach led the field in exploiting the value of the ageless registration numbers on pre-suffix Routemasters, with re-registering taking place from 1986 onwards. Routemaster services linked central Glasgow with the vast housing schemes at Easterhouse and Castlemilk, and re-registered RM 1599 is seen in a typical Glasgow setting at St Enoch about to depart for the latter. Ken Blacker

Three Scottish Bus Group subsidiaries pinned their faith on the Routemaster as an easily available, low cost tool for keeping predators at bay after deregulation, and for attacking competitors such as Strathclyde Buses. Clydeside rapidly built up a substantial fleet which initially ran mainly on local services in the Paisley area where RM 2083 is seen, although their ultimate task was mainly to spearhead the defence of the highly profitable Paisley to Glasgow corridor. Clydeside's Routemaster fleet developed into the largest outside London and was distributed over four depots. Their Routemasters, like so many others that were to follow, carried full Johnston-style destination displays. Male names from Rambo to Rumplestilkskin were given to Clydeside's Routemasters and carried on the buses as shown. Steve Fennell/John Fozard

By no means all operators who sampled the Routemaster were sufficiently smitten to purchase any for themselves. The cost of reverting back to crew operation had to be balanced against an unquantifiable upsurge in trade (if any), and the conclusion was often that the risk could not be justified. The vehicles themselves, elderly and totally non-standard, may not fit satisfactorily within existing maintenance schedules and, no matter how well maintained they may currently be, the procurement of spare parts could be expected to pose a problem for the future. Sometimes these considerations were all subordinated to the need for a new image to deter, fight off or even spearhead competition, but such was not the case with Brighton & Hove in whose service RM 1721 is seen on 31st January 1987. On loan from London Bus Sales and still in London red, albeit with local fleet name and posters, this vehicle did not find a new home and was sold just over a year later for scrap.
A.E. Hall, courtesy Alan B. Cross

The south of England's first bus war after deregulation took place in Southampton, sparking off the introduction by Southampton City Bus of Routemasters on new route 16 on 29th May 1987 under the banner "New Buses! New Times! Change Given!" Twelve Leyland-engined RMs were acquired to cover a six bus schedule, the balance being employed when crew operation was extended to two existing services in August. Southampton's red and cream livery is demonstrated by RM 1793 photographed at Redbridge Hill in July 1987 exhorting passengers to "Hop on at the back", though later acquisitions carried far smaller fleet names. Mechanical maintenance on the RMs was kept only at minimum level, resulting in the requirement to purchase further vehicles to replace defective ones from the original batch. Interior cleaning also appeared to be only cursory in nature. With the competitive threat much diminished through stabilisation of services in the area, the pioneer RM route 16 was withdrawn in January 1988. The RMs, latterly on routes 17 and 17A, last ran on 14th January 1989, and only a few were still serviceable at the end. R.J. Waterhouse

The Routemaster revival made an innocuous start when five vehicles were sold in January 1985 by London Bus Sales to Stagecoach Ltd, a small Perth operator who used them, still in London livery, on local operations in Perth and Dundee. This was, of course, well ahead of Deregulation Day, although in bus headquarters throughout the country plans were secretly being hatched in readiness.

Scotland figured again during 1985 when, in July of that year, RM 652 was borrowed by Clydeside Scottish Omnibuses Ltd for evaluation from each of its four garages. Purchased in August, and shortly afterwards repainted into Clydeside livery, it was soon joined by RM 694 (technically initially on loan) and these two became the precursors of large fleets of Routemasters purchased by Clydeside, and also by fellow Scottish Bus Group subsidiaries Kelvin and Strathtay. Clydeside took its Routemasters so seriously that it even published a 'Rodney the Routemaster' booklet for younger travellers and painted former RM

652 to match. Viewing the reintroduction of crew operation as an effective weapon with which to combat deregulation, these three operators made Scotland the focal point for the Routemaster comeback. The impact was heightened on Deregulation Day by Stagecoach whose invasion of Glasgow by Routemasters in striking white livery with orange, red and blue reliefs and the faintly ridiculous title of Magicbus, heralded the first major expansion of a company which was soon to cannily purchase for itself a position of power in the bus industry as a consequence of privatisation.

Flirtation with the Routemaster spread far and wide. Stagecoach subsidiaries Cumberland, United Counties and East Midland were obvious candidates, each claiming that the use of Routemasters was its own decision and not dictated by Stagecoach headquarters. Others, such as Blackpool Transport, Burnley & Pendle and Southend Transport became staunch RM fans. Greater Manchester's Routemaster

operation was interesting in following on a quarter of a century after RM 1414 had been tried out in the city only to be firmly rejected at the time. East Yorkshire's Routemasters were splendid in reviving the glorious indigo and primrose with white roof band of happier, pre-NBC days. Some small operators who tackled deregulation with Routemasters such as Coster's Citilink of Hull, M.W. Gagg of Bunny near Nottingham, Blue Bus Services of Eccles and Dorset-based Verwood Transport (the latter using RMAs) found the pace too much. Even some of the larger Routemaster operators, notably Kelvin, bit off more than they could chew and had to reduce or, in the case of Southampton Citybus, return to conventional vehicles. Others, such as Maidstone Borough, Provincial, Brighton & Hove and Northern General, toyed with the idea of running RMs but thought better of it, whilst a substantial 'Red Admiral' scheme at Portsmouth for which RMs were acquired failed to get off the ground.

Of particular interest in the continuing story of the Routemaster has been the significant role played by the staff of London Bus Sales in attempting to foster large-scale overseas markets for redundant vehicles. In the first scheme a possible market for a huge number of RMs was thought likely to exist in China where public transport in some major cities was under considerable pressure which could be eased through the use of double deckers. Positive steps were taken to exploit this market beginning with the overhaul at Aldenham in the summer of 1984 of RM 1288 which included reversing the staircase and rear platform for right hand running (reminiscent of a similar 1958 conversion of RTL 3), the fitting of sliding windows, and repainting in yellow livery. Accompanied by RM 1873, still in London condition, RM 1288 set sail early in September heading for the main premises in Hong Kong of China Motor Bus who were to act in the capacity of dealers for evaluation trials. Early tasks for the two buses were scheduled at a couple of trade fairs in Hong Kong where one was used as an exhibit and the other to ferry underprivileged children, the elderly and handicapped visiting the exhibits. Afterwards RM 1288 crossed into mainland China for a two month trial in Shenzhen, the centre of a Special Economic Zone just across the border. Modifications carried out by Citybus (Hong Kong) saw RM 1288 given a front entrance/exit just behind the driver with power-operated doors both here and on the rear platform. RM 1873 was converted to bring it in the same condition and in this form the two vehicles held 131 passengers – 36 seated upstairs, 28 seated downstairs and 67 standing! The first real test came when RM 1288 worked in service conditions in Gangzhou (formerly known as Canton) where it reportedly caused quite a stir and was received with spontaneous hand clapping. Some 70,000 miles of virtually trouble free running and its general warm reception gave London Transport the confidence to hit the headlines in January 1986 with the news that negotiations were taking place for a multi-million pound contract to sell up to 1,300 Routemasters to China for operation in Gangzhou and also in the capital, Beijing (Peking). The Chinese authorities reputedly wished to barter uniforms in part payment for the £20m deal. Next stage in the saga was for RM 1873 to make a 2,000 mile overland promotional journey to Beijing starting at the end of January and ending up back at Hong Kong some ten days later. Alas, all then went quiet and other influences began to bear which ensured that the deal did not go through. After a lengthy period of storage RMs 1288/1873 were sold to Citybus (Hong Kong) in September 1988.

Top **Carrying high hopes for a volume sale of RMs to China, which never materialised, RM 1288 arrives at Hong Kong at the start of its career in the Far East.** John Laker

Centre **After receiving a livery modification and local lettering, RM 1288 attracts attention in Gangzhou when a myriad bicycles typify the urban Chinese scene.** John Laker

Bottom **RM 1873 after conversion to dual-door mode, photographed prior to its long journey to Beijing.** John Laker

United Counties announced its Routemasters as being on a nine months trial when they commenced local operation in Bedford on 1st February 1988, with Corby added two months later. A trial it may have been, but success must be presumed as operation continued. The company's reconversion to crew operation in towns which would not perhaps normally be judged as offering the highest potential, was accompanied by a high key publicity drive featuring prominently the name of the bus itself, ROUTEMASTER. The company's attractive green livery, brightened by bands of orange, yellow and cream, carried the word ROUTEMASTER prominently on side panels, front dome and bonnet top. This theme was further promoted in publicity literature and it also appeared on the route diagram carried on the main side panels of each vehicle. RM 528 is seen at Putnoe on Bedford cross-town route 101. Mike Harris

Burnley & Pendle Transport, faced with unwelcome competition on its main Burnley to Colne corridor, responded with Routemasters on 14th March 1988, its first rear loaders for sixteen years. Three of the initial small fleet of four were rushed into service in London colours but all were soon repainted in the fleet livery of red and cream. Operating a 20-minute frequency just in front of the competing vehicle, they exploited their London origin to the full in being marketed as EastEnders after the popular television soap opera of the same name. To further enhance public awareness, each vehicle was prominently named after a well known character in the show, the name being displayed in full view on the front advert panels. For the East Enders programme itself a Routemaster was considered a useful adjunct as representing a typical London bus, and the role was given in January 1988 to RM 318 on long term loan from London Buses. Mike Harris

Unlike other purchasers of secondhand Routemasters, Blackpool Transport still possessed a small fleet of its own open rear platform double deckers so the Routemasters were perhaps less of a novelty than was the case elsewhere. With the exception of the London fleet, buses of this configuration were extinct in most other parts of the country by January 1986 when RM 1583 was received at Blackpool for evaluation, subsequently forming the nucleus of a small fleet from April onwards. When finally painted in Blackpool colours these vehicles looked very fine in a 'vintage' livery of red and white with black edging and prominent lining-out on the lower panels, although a later 1988-acquired batch were more sombrely turned out with most of the frills omitted. All of Blackpool's RMs were Leyland engined. RMs 1640 and 1735 are seen in their first summer before the class was transferred to work the main road to St Annes. The second, 1988 batch was acquired to fight off a Fylde Borough Transport attack along Blackpool's famous sea-front tramway, resulting in RMs further depleting tramcar receipts. John Fozard

The competitive threat in Hull posed by a number of independent entrepreneurs caused one of the two main incumbent operators, East Yorkshire, to turn to Routemasters to protect itself on routes 56 and 56A where an improved 10-minute interval was provided over the long common section. The operation was marketed as the East Yorkshire Clipper although this name was not carried on the vehicles, and the claim was made that it would "bring back the good old days of clippies". The faster running time with which Routemasters could be scheduled compared with opo double deckers enabled a vehicle saving to be made, but even so the company was quick to admit that crew operation could only just about be justified financially. In total contrast to the state-of-the-art colour schemes favoured by several operators, East Yorkshire chose to turn the clock back more than half a century by reviving the classic blue livery which had made its fleet so distinctive in pre-NBC days. The seven RMs with which operation commenced on 3rd May 1988 looked superb as witnessed by RM 1366 passing through Longhill during its second week of operation. R.J. Waterhouse

Southend Transport, which unlike most operators had never completely abandoned conductors, borrowed Routemasters on occasions prior to placing its own into service on 3rd September 1988 in a rolling programme embracing three of the undertaking's main routes. Emerging from a financial and managerial crisis, the company adopted a new image with the arrival of its Routemasters based on a livery of blue and white with red band, new style fleet name and, unfortunately, most unattractive stick-on fleet numbers. Before entering service the vehicles were renovated inside as well as out, and were provided with white ceilings and up to date lighting. Simon Fozard

Three months after coming into Stagecoach ownership, Cumberland began Routemaster operation of cross-town Carlisle service 61 on 26th October 1987. A special Routemaster version of the post-NBC red and sandstone livery was applied to eight vehicles, six of which were ex-Kelvin plus two purchased direct from London. One of the latter, RM 2024, is seen in Stonegarth. The allocation of RMs and DMSs (the latter transferred in from Hampshire Bus) did not pass without incurring strong public protest over the number of elderly buses transferred into Carlisle following the Stagecoach takeover. RM 2024 has a mileage recorder fitted to its back hub. R.J. Waterhouse

Having purchased East Midland and Mansfield & District from their original management buy-out team, the Stagecoach group lost little time in transferring United Counties RM 980 from Corby for experimental operation in its newly acquired territory. The handsome two tone green and cream Mansfield & District livery suited it well although its use as late as May 1989 was unexpected as the Stagecoach corporate livery policy was now in force. Despite the London sounding name on the signpost, RM 980 is storming uphill from Whatstandwell to Crich in rural Derbyshire on a Sunday service linking Shirebrook and Matlock with the Crich tramway museum. Bob Pennyfather

The first old established independent operator to purchase Routemasters was W. Gash & Sons Ltd, who rushed two into service at deregulation on a local run in Newark. The company's RM fleet eventually stood at three. Long renowned for its immaculate fleet, Gash flew headlong into deregulated expansion and eventually went far out of its depth. The RMs ran in as-received condition, dents and all; only one received Gash's own distinctive green livery, and this was right at the end, after the company had been sold to Yorkshire Traction and placed under operational control of Lincolnshire Road Car. Sadly the business ceased without warning in May 1989 after a Department of Transport examination which resulted in revocation of its operating licence. In that month, the repainted former RM 757 is seen in Newark. Simon Fozard

A line up of Routemasters beneath the coconut palms of rural Sri Lanka finds former RMs 283, 1067, 1294 and 1413 at Negombo depot. By far the most complex buses ever to run on the island, they are already showing maintenance difficulties.
Peter McMahon

During 1985 rumours were rife that up to 400 RMs could be destined for export to Sri Lanka, and credibility was lent to this rumour when, in December, RM 499 was despatched to the Indian Ocean island. Earlier in the year RM 499 had earned notoriety when it was converted to drive from upstairs for a television advertisement portraying a driverless bus taking a fare dodger to Wormwood Scrubs prison. The driver was, in fact, hidden from view on the upper deck, looking through the blindbox and operating the bus with a special throttle and hand-worked air brake, chain and sprockets and an extra long shaft being used to connect to the existing steering wheel. Needless to say this ingenious conversion, carried out at Aldenham, was removed prior to sale. (Another withdrawn RM, 1066, was similarly converted for the première of the West End show, 'Phantom of the Opera' in November 1986). Sri Lanka is, of course, well

remembered under its old title of Ceylon as having been the final home for no fewer than 1,273 RTs, RTLs and RTWs purchased from London Transport between 1958 and 1968. These had operated, often in very decrepit condition, in the Colombo area though hardly any still remained by 1985. After a period of gestation, RM 499 was followed into the fleet of the Sri Lanka Central Transport Board by a further forty RMs purchased by the Crown Agents on behalf of the Overseas Development Administration as part of a British Government rehabilitation aid scheme. The forty vehicles were all fitted at Chiswick with improved air filters and full depth sliding windows in the main side apertures on both decks (leaving the front quarter-drops in place), and a cast metal plate on the staircase panel reading 'Provided under British Aid'. Amid a welter of press and television coverage the fleet was loaded at Sheerness docks aboard the

Liberian-registered *Nasac Takara* on Friday 16th December 1988 and set sail the same day, their future perhaps a little uncertain in an island wracked by racial tension.

On 1st July 1988 a Reuter report from Tokyo broke the news that thirty Routemasters were to be purchased by the Nissho Iwai Corporation within a year for conversion into restaurants, noodle shops and bars. Firm plans at the time encompassed six buses of which the first was due to be reincarnated in the same month as a bar in the city of Osaka. The other five, would be based in Tokyo, Nagoya and ski resorts in the Nagano Prefecture. This venture however fell through, but not before a few vehicles reached Japan.

By April 1989 disposals had reduced the number of standard RMs still owned by London Buses to fewer than 500, with only 39 RMCs, 11 RCLs and 15 RMAs still on the Company's books. By no means all of those were still in use. However, the RML class continued to rule supreme. Apart from RML 2557, lost as a result of an arson attack at Holloway in February 1983, only RML 900 had been sold. Disposed of in February 1988 as being beyond economic repair after colliding head-on with a mechanical digger in July 1987, it was purchased by Clydeside who towed it home and quickly decided that repairs were possible after all! RML 900 was thus the only RML in provincial service, having commenced operations with Clydeside on 4th June 1988, even appearing back in London for a guest visit on Finchley garage routes 13 and 26 (the latter a one-man operation since September 1980) on 18th June 1988.

Clydeside gained renown in June 1988 by placing RML 900 into passenger service, the only one of its class outside London. Repair work to correct the extensive frontal damage which had led to its sale resulted in the fitting of a fixed windscreen and DMS-type front trafficators, and it was also extensively refurbished internally including the provision of fluorescent lighting. In 1988 the large Clydeside Routemaster fleet was riding high; some RMAs had been purchased and the RMs enjoyed a high profile in Glasgow, disguising a mounting disquiet over the company's perturbing financial performance. RML 900's service debut was actually back in London where, on 18th June 1988, it ran a couple of journeys from its original garage, Finchley, first on route 13 and then on omo route 26. The vehicle featured a smiling 'Oor Wullie', a favourite cartoon character with readers of the Scottish Sunday Post.
R.J. Waterhouse

Very severe competition resulting from deregulation was the spur for Greater Manchester Buses to turn to Routemasters as a retaliatory tool. Starting on 5th September 1988 a new route numbered 143 was superimposed upon the proliferation of buses already serving the Wilmslow Road and was unusual in extending RM operation right through to late evening whereas most other operators chose not to run them after the evening peak. Aptly called the Piccadilly Line after its city terminus at Piccadilly bus station, route 143 deftly followed a London theme with retention of full London style livery even to the use of a symbol loosely resembling the famous London roundel. The London fleet numbers were retained in their usual position although local ones were also allocated from 2200 upwards, RM 2200 being unique amongst the ten vehicles in that its London and GMB numbers coincided. R.J. Waterhouse

With forthcoming privatisation of the Scottish Bus Group high on the agenda in 1989, the need was seen to stabilise the finances of its less healthy subsidiaries. As a result Clydeside was merged into the much older established Western Scottish, from which it had been an offshoot in the first place at an earlier reorganisation. The Western management saw no place in its scheme of things for the Routemasters, and in August 1989 it was announced that the 70-strong fleet along with 130 conductors would be made redundant. In the company's words, the Routemasters had served their short term competitive purpose. In a sight soon to be no more, four RMs are seen in Glasgow's Buchanan Street bus station with one of the soon to be unemployed conductors assisting RM 391 to reverse out. The Western fleet numbers in the red central band are an outward sign of new ownership.

Almost three years into deregulation, Stagecoach decided to launch head on into a battle with Strathtay in both Perth and Dundee. Part of the Strathtay response was the adoption of a new 'old' image for its Perth based RM fleet coupled with an extension of their sphere of operation within the city. The deep red and cream livery, maintained for many years by predecessor company Walter Alexander & Sons Ltd in memory of the Perth municipal operations of long ago, was revived under the guise of Perth City Transport. RM 1821, photographed in October 1989, displays its newly acquired colours which suit it admirably and, being redolent of an era long past, are as different as could possibly be from the totally modern stripey orange, blue and white carried previously. John Fozard

APPENDIX 1
ROUTEMASTER REPLACEMENT DESIGN WORK, 1989/90

By the mid-1980s it had become clear that no in-house design initiatives on the scale that produced the Routemaster would be possible in future. The dissipation on economy grounds of the Company's design expertise, including closure of the famous Chiswick experimental shop, meant that the capability was no longer there. When, at the end of the decade, London Buses' management decided to test the feasibility of producing a new bus in the same mould as the Routemaster it had to turn to manufacturers for design and development work. Almost unbelievably, more than twenty years after the last old style double decker with front engine, half width cab and open rear platform had been built for use in the United Kingdom, London Buses asked for just such a vehicle to be put on the drawing board.

Dennis showed interest in developing a chassis design for a 9.9 metre long vehicle whilst, during 1990, Alexander and Northern Counties produced drawings for competing body designs, both of which followed the current idiom in being rather angular. Slightly longer than the RML, the complete vehicle would have seated either 79 passengers or 74 if a straight staircase had been specified. Undoubtedly such a vehicle, if put into production, would have been expensive to produce because of high setting up costs in relation to the comparatively low volume required, and it was perhaps no surprise that London Buses, underfunded and facing uncertainty as to its future direction, opted instead to embark on the much cheaper alternative of a modernisation programme for the RML fleet.

Drawings produced by Northern Counties are reproduced below. The alternative rear end layouts include one with improved access for the disabled, allowed for also in the Alexander designs shown opposite.

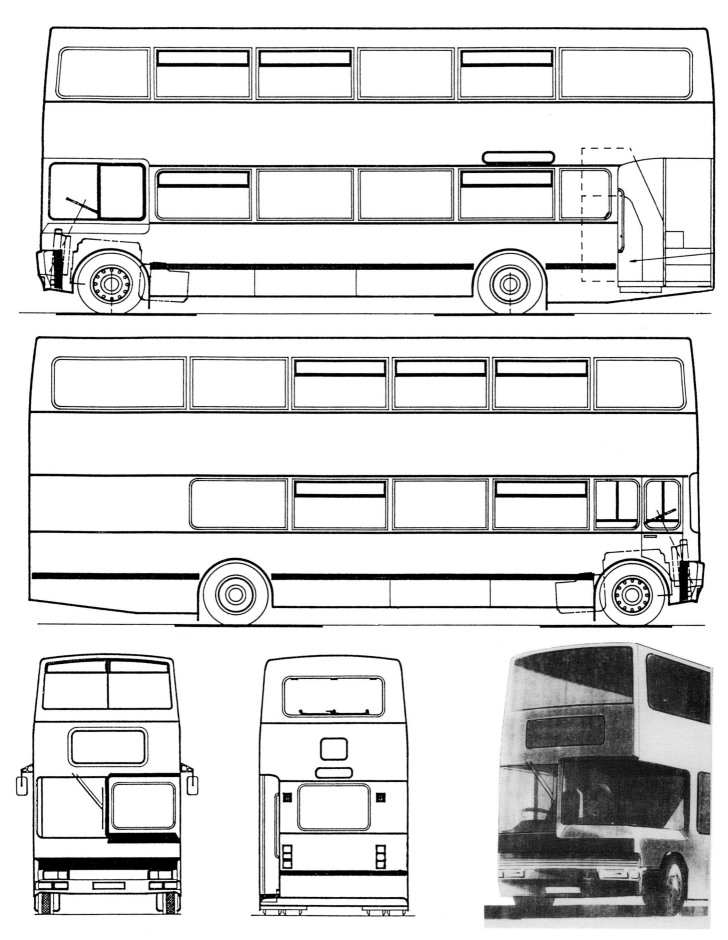

APPENDIX 2
OVERSEAS GOODWILL TRIPS BY ROUTEMASTERS, 1970-1989

Routemaster production having ceased in 1968, goodwill visits overseas could no longer call upon the services of brand new traditional open-platform London double deckers. Faced with the choice of sending new DMSs or overhauled Routemasters, a decision was made in favour of the latter. Until 1973, each vehicle overhauled for an overseas trip retained the same body with which it entered the works, but this practice fell by the wayside afterwards.

Destination	Vehicle(s)	Notes
Hamburg	RM 403/404	Trade week, September 1971
Stuttgart	RML 2266	British week, October 1972
Pittsburgh	RML 2261	October 1972 (last trip across the Atlantic)
West Berlin	RML 2462	Travel conference, April 1973
European Tour	RML 2536	Belgium, Holland, West Germany, Switzerland, France, Sept/Oct 1973
Milan	RML 2536	British week, November 1973
Munich	RML 2533	Trade fair/British week, November 1973
Boulogne	RM 1359	Boulogne Chamber of Trade all-over advert livery, May 1974
Paris	RML 2728/2747	For new Marks & Spencer store, February 1975
Brussels	RML 2728/2747	For new Marks & Spencer store, March 1975
Douai	RML 2404	October 1979
Holland	RML 903	September 1987, Leaside District initiative
Paris	RM 14/1185	For Champs Elysees Grand Parade as part of the French Revolution bi-centenary celebrations

APPENDIX 3
ALLOCATION OF COUNTRY BUS AND GREEN LINE ROUTEMASTERS AT 1st JANUARY 1970

Addlestone (WY)	RMC 1471, 1473, 1489, 1491, 1494, 1504, 1513, 1518, 1519	Total 9
Dunton Green (DG)	RCL 2235, 2236, 2238-2242, 2247, 2251-53	Total 11
East Grinstead (EG)	RML 2306	Total 1
Garston (GR)	RML 2318, 2320, 2419-25, 2427-35, 2437-40	Total 22
Godstone (GD)	RCL 2226, 2237, 2250	Total 30
	RML 2307, 2310-17, 2319, 2329-36, 2346-54	
Grays (GY)	RMC 1462, 1463, 1466, 1468, 1475, 1476, 1492, 1496, 1497, 1502, 1515	Total 17
	RCL 2228, 2234, 2243, 2245, 2246, 2248	
Guildford (GF)	RMC 1457, 1458, 1465, 1478, 1484, 1488, 1490, 1499	Total 8
Harlow (HA)	RMC 1472, 1477, 1479, 1480, 1483	Total 18
	RML 2309, 2413, 2442, 2444, 2445, 2448-50, 2453, 2456, 2458-60	
Hatfield (HF)	RMC 4, 1469, 1482, 1487, 1495, 1498, 1500, 1505, 1506, 1508-10, 1512, 1516, 1520	Total 15
Hemel Hempstead (HH)	RML 2414, 2457	Total 2
Hertford (HG)	RMC 1453, 1454, 1459-61, 1467, 1470, 1474, 1503	Total 9
High Wycombe (HE)	RML 2411, 2412, 2415-18	Total 6
Northfleet (NF)	RML 2322-28, 2337-45, 2355, 2426, 2446, 2447, 2451, 2452, 2454, 2455	Total 24
Reigate (RG)	RML 2308	Total 1
Romford (RE)	RCL 2218-25, 2227, 2229-33, 2244	Total 15
Stevenage (SV)	RMC 1501, 1507, 1511, 1514, 1517	Total 5
Windsor (WR)	RMC 1455, 1456, 1481, 1485, 1486, 1493	Total 15
	RCL 2249, 2254-60	
	RML 2436	

APPENDIX 4
ROUTES SCHEDULED FOR ROUTEMASTER OPERATION (CENTRAL BUSES), 1959-1989

Appendix compiled by Peter Nichols and Les Stitston. The photographs accompanying this appendix are by Colin Routh (RM 1380 and others in Parliament Street), Colin Fradd (RM 487 in Whitehall, RM 874 at Finsbury Park, RM 315 at Erith, RCL 2251 at Trafalgar Square), Colin Stannard (RM 926 at Streatham Vale), Capital Transport (RM 2040 at Harrow), Peter J. Relf (RM 1842 at Hounslow Heath).

Route	Garage	F.D.O.	L.D.O.	Notes
				Sunday allocation on daily route implies that while this garage worked RMs on Sundays on this route it also worked other types on other days.
				Sunday only allocation on daily route implies that this garage only worked on this route on Sundays using RMs. Other garages worked on other days.
				References T/B4-T/B14 indicate trolleybus to diesel conversion stages
1	NX	22.03.69	05.06.87	Saturday allocation on Monday to Saturday only route. Monday to Saturday from 23.02.75. Daily from 28.10.78
	TL	27.01.75	01.11.85	Monday to Friday only allocation, Monday to Saturday from 31.10.81
1A	NX	25.10.69	28.10.78	Sunday only route
		20.04.84	21.10.84	Sunday only route
	TL	20.04.84	21.10.84	Sunday only route.
2	W	08.02.56	08.08.56	RM 1 trial service.
		20.08.58	03.08.59	RM 1 trial service.
		15.06.59	10.11.59	RMs 10, 19, 30, 35, 38, 39, 41, 42, trial service. Daily.
		29.12.62	12.06.70	Saturday and Sunday allocation on daily route. Daily from 17.06.67
	SW	10.07.66	20.06.86	Sunday allocation on daily route. Daily from 01.05.67
	GM	03.01.71	22.10.78	Sunday only allocation on daily route.
		29.10.84	26.04.85	Monday to Friday only allocation on daily route.
	N	29.10.78	19.04.81	Sunday only allocation on daily route.

Route	Garage	F.D.O.	L.D.O.	Notes
2A	SW	01.05.67	04.01.74	Monday to Friday only route.
2B	N	01.11.64	24.04.81	Sunday allocation on daily route, Saturday/Sunday from 23.01.66. Daily from 01.11.66.
	SW	01.11.66	23.05.87	Daily allocation. Monday to Saturday only allocation on daily route from 08.01.72, Monday to Friday only from 17.06.72, Sunday to Friday only from 28.10.78. Daily from 25.04.81. Saturday only from 27.10.84, Sunday only from 07.02.87.
	FY	14.06.70	22.10.78	Sunday only allocation on daily route.
	GM	17.06.72	26.10.84	Saturday only allocation on daily route, Saturday and Sunday only from 28.10.78. Daily from 25.04.81, Monday to Friday only from 25.06.83.
3	N	01.11.64	22.04.81	Sunday allocation on daily route. Daily from 01.12.64.
	CF	22.12.64	20.06.86	Daily allocation. Monday to Saturday from 19.10.75, daily from 22.04.79. Sunday only allocation from 26.04.81, Sunday to Friday only from 05.09.82. Daily from 27.10.84.
	SW	25.04.81	03.09.82	Daily allocation.
		23.06.86	06.02.87	Monday to Friday only allocation on daily route.
	Q	25.04.81	03.08.85	Monday to Saturday only allocation on daily route. Daily from 04.09.82, Saturday only from 27.10.84.
		02.11.85		Daily. Monday to Saturday from 11.07.87.
	WL	28.10.84	01.11.85	Sunday to Friday only allocation on daily route. Daily from 03.08.85.
4	HT	04.09.71	01.04.72	Saturday allocation on Monday to Saturday route.
		15.07.72	18.01.75	Saturday allocation on Monday to Saturday route.
		31.10.81	26.01.85	Monday to Saturday allocation on Monday to Saturday route.
5	PR	11.11.59	26.01.65	Daily allocation (T/B 4). Monday to Saturday only from 09.05.62, Monday to Friday only from 15.08.62.
		08.09.68	16.04.71	Sunday only allocation on daily route.
	WH	11.11.59	16.04.71	Sunday to Friday only allocation on daily route (T/B 4). Monday to Friday only from 09.05.62, Monday to Saturday only from 07.09.68.
		25.04.81	02.08.85	Monday to Saturday only allocation on daily route. Monday to Friday only from 06.09.82.
	U	04.09.82	01.11.85	Daily allocation.
5A	PR	11.11.59	30.12.66	Monday to Friday only route. (T/B 4)
	WH	02.01.67	06.09.68	Monday to Friday only route.
5B	PR	13.05.62	01.09.68	Sunday only route. Saturday and Sunday only route from 18.08.62. Sunday only route from 30.01.65.
	WH	13.05.62	07.10.62	Sunday only route. Sunday only allocation on Saturday and Sunday route from 19.08.62.
		31.01.65	01.09.68	Sunday only route.
5C	PR	30.01.65	31.08.68	Saturday only route.
	WH	30.01.65	31.08.68	Saturday only route.
6	AC	17.02.65		Daily allocation. Monday to Saturday from 06.06.87.
	H	21.04.65	24.04.81	Daily allocation.
	AG	25.04.81	22.11.91	Daily allocation. Monday to Saturday from 06.06.87.
	BW	23.11.91		Monday to Saturday allocation on daily route.
6A	H	21.04.65	17.01.70	Monday to Saturday only route. Saturday only route from 07.09.68.
6B	WW	01.02.64	31.08.68	Saturday only route.
	AC	20.03.65	31.08.68	Saturday only route.

Route	Garage	F.D.O.	L.D.O.	Notes
7	X	02.12.63		Monday to Saturday only route. Daily from 25.04.81. Monday to Saturday allocation on daily route from 01.02.86.
8	AC	22.01.58	31.10.59	RML 3 trial service.
		04.06.59	10.11.59	RMs 5, 7, 22, 24, 25, 28, 86 trial service.
		17.02.65	03.09.82	Daily allocation.
		23.11.91		Monday to Saturday allocation on daily route.
	BW	13.01.65		Daily allocation. Monday to Saturday only allocation on daily route from 16.01.88.
8A	BW	13.01.65	20.06.86	Daily route. Monday to Friday only route from 25.10.69.
8B	AC	26.01.58	25.10.59	(Old Ford — Alperton) Sunday only route. RML 3 trial service.
		07.06.59	08.11.59	Sunday only route. RMs 5, 7, 22, 24, 25, 28, 86 trial service.
	W	07.09.68	12.06.70	(Cricklewood — Bloomsbury) Monday to Friday only route.
9	PR	15.11.59	01.09.68	Sunday only allocation on daily route. (T/B 4)
		21.12.68	10.01.70	Sunday only allocation on daily route.
	M	05.04.63	24.06.83	Daily allocation.
	D	17.05.63	27.10.78	Monday to Saturday only allocation on daily route. Monday to Friday only from 18.04.70, Sunday to Friday only from 16.01.71. Monday to Friday only allocation on Monday to Saturday route from 09.04.71.
		31.01.81	24.04.81	Monday to Saturday only route.
	R	28.10.78	30.01.81	Daily route.
	AG	25.04.81	14.08.87	Daily route.
	V	25.06.83	05.01.90	Daily route. Monday to Saturday only allocation from 27.04.85, daily from 25.10.86. Monday to Saturday allocation on daily route from 21.11.87.
	S	28.04.85	19.10.86	Sunday only allocation on daily route.
		15.04.89		Monday to Saturday only allocation on daily route.
	GM	15.08.87	14.04.89	Daily allocation. Monday to Saturday from 21.11.87.
9A	D	09.04.71	22.10.78	Sunday only route.
		01.02.81	19.04.81	Sunday only route.
	M	09.04.71	22.10.78	Sunday only route.
		01.02.81	19.04.81	Sunday only route.
10	BW	17.01.65	27.10.72	(Victoria — Abridge) Sunday allocation on daily route. Saturday and Sunday from 01.11.69.
	S	13.08.88		(Hammersmith — Kings Cross) Monday to Saturday allocation on daily route.
	GM	15.08.88	14.04.89	Monday to Friday only allocation on daily route.
	HT	15.04.89		Monday to Saturday only allocation on daily route.
11	R	05.06.59	10.11.59	RMs 14, 29, 31-34, 37, 52 trial service.
		03.10.65	24.06.83	Saturday and Sunday allocation on daily route. Daily from 01.02.66.
	GM	01.11.64	27.12.70	Saturday and Sunday only allocation on daily route. Sunday only from 24.01.70.
	GM	25.06.83		Daily allocation. Monday to Saturday from 26.03.88.
	D	01.02.66	24.04.81	Daily allocation.
	AG	25.04.81	14.08.87	Daily allocation. Monday to Saturday only allocation on daily route from 02.11.85.
12	S	21.12.68	30.10.81	Saturday allocation on daily route. Monday to Saturday only allocation on daily route from 12.05.73.
		27.10.84	24.10.86	Daily allocation.
		13.08.88		Monday to Saturday only allocation on daily route.
	PM	02.09.72		Saturday and Sunday allocation on daily route. Daily from 24.03.73. Monday to Saturday from 15.10.77, daily from 04.09.82. Monday to Saturday from 13.08.88.

Route	Garage	F.D.O.	L.D.O.	Notes
	WL	03.09.72	01.11.85	Sunday allocation on daily route. Daily from 06.01.73, Monday to Saturday from 23.02.75, daily from 28.10.78. Monday to Saturday only allocation from 27.10.84.
	ED	10.03.73	24.10.86	Daily allocation.
	Q	04.11.85		Monday to Friday only allocation on daily route. Daily from 25.10.86, Monday to Saturday from 13.08.88.
13	AE	17.12.62	27.10.78	Monday to Saturday only route.
	RL	17.12.62	21.03.69	Monday to Friday only allocation on Monday to Saturday route.
	Q	24.03.69	23.01.70	Monday to Friday only allocation on Monday to Saturday route.
	MH	26.01.70	27.10.78	Monday to Friday only allocation on Monday to Saturday route.
	FY	28.10.78		Daily allocation. Monday to Saturday from 07.02.87.
14	NB	12.05.63	27.12.70	Sunday only allocation on daily route.
	AF	02.10.63		Daily allocation. Monday to Saturday from 02.02.87.
	J	17.10.63	03.09.71	Daily allocation.
	HT	04.09.71	06.02.87	Daily allocation.
15	X	07.12.63	28.08.82	Saturday and Sunday only allocation on daily route. Saturday only from 25.04.81.
	X	04.03.89		Monday to Friday only allocation on daily route.
	U	01.04.64		Daily allocation. Monday to Saturday from 06.06.87.
15A	U	20.05.85	05.06.87	Monday to Friday only route.
15B	U	06.03.89		Monday to Friday only route.
X15	U	06.03.89		Monday to Friday only route. RMC.
16	W	24.12.62	14.12.73	Daily allocation.
		24.05.80	20.11.87	Daily allocation.
16A	W	24.05.80	20.11.87	Monday to Saturday only route.
17	HT	01.02.61	24.01.75	Daily allocation (T/B 9). Monday to Friday only route from 16.01.71.
	WL	01.01.67	10.01.71	Sunday only allocation on daily route.
18	X	03.01.62	06.09.68	Daily allocation (T/B 13). Monday to Saturday only from 31.12.66.
	SE	03.01.62	21.04.79	Daily allocation (T/B 13).
	ON	11.03.64	06.05.79	Sunday to Friday only allocation on daily route. Sunday only from 28.10.72.
18A	X	19.06.67	03.09.82	Monday to Friday only route. Officially converted to DM from 17.04.81 but was mainly RM in practice.
19	B	16.07.72	01.11.85	Daily allocation.
	HT	16.07.72	14.04.89	Daily allocation. Monday to Saturday from 06.07.87.
	GM	02.11.85		Monday to Saturday only allocation on daily route.
21	SP	01.03.75	31.01.86	Daily allocation. Monday to Saturday from 02.06.84.
	NX	01.03.75	31.01.86	Monday to Saturday only allocation on daily route. Monday to Friday only from 04.09.82.
22	H	*24.06.59	10.11.59	RMs 18, 40, 43, 47, 54, 55, 99 trial service. *Probable date.
		25.04.65	06.09.68	Sunday allocation on daily route. Daily from 02.10.67.
	B	11.11.67	01.11.85	Daily allocation.
	CT	07.09.68	10.07.87	Daily allocation. Monday to Saturday from 07.02.87.
	AG	11.07.87	20.11.87	Monday to Saturday only allocation on daily route.
	AF	11.07.87		Monday to Saturday allocation on daily route.
	GM	27.05.91		Monday to Friday only allocation on daily route.
23	PR	11.11.59	02.06.84	Monday to Saturday only route (T/B 4). Daily from 07.09.68. Sunday only allocation on daily route from 25.10.69, Saturday only from 18.04.70. Saturday and Sunday only allocation from 30.10.71, Saturday only from 05.01.74.
	BK	03.03.64	17.07.70	Monday to Friday only allocation on Monday to Saturday route. Sunday to Friday only allocation on daily route from 07.09.68.
	U	07.09.68	17.05.85	Saturday only allocation on daily route. Sunday only from 18.04.70. Sunday to Friday only from 18.07.70. Monday to Friday from 20.04.74. Daily from 04.09.82.
	WH	25.10.69	03.09.82	Monday to Saturday only allocation on daily route. Daily from 05.01.74, Monday to Saturday only from 20.04.74. Daily from 25.04.81.
24	CF	07.11.63	06.11.65	Daily allocation.
		12.06.66	18.10.75	Daily allocation.
		22.04.79	24.10.86	Daily allocation.
	GM	24.01.70	10.06.72	Saturday only allocation on daily route.
25	WH	06.02.66	01.11.85	Sunday allocation on daily route. Saturday and Sunday from 25.10.69, Sunday from 19.04.70. Daily from 08.01.72, Monday to Saturday from 03.08.85.
	BW	26.10.69	15.01.88	Sunday allocation on daily route. Daily from 08.01.72, Monday to Saturday only allocation from 02.02.85.
26	FY	13.06.70	26.09.80	Monday to Saturday only route. Daily from 28.10.78.
	W	15.06.70	15.01.71	Monday to Friday only allocation on Monday to Saturday route.
27	V	16.08.59	08.11.59	RMs 36, 57, 75, 89, 94, 95, 96, 101, 111 trial service.
		19.04.70	18.07.86	Sunday allocation on daily route. Sunday to Friday only allocation from 13.06.70, daily from 28.10.78.
	R	06.02.66	21.10.78	Sunday allocation on daily route. Saturday and Sunday only allocation on daily route from 18.04.70, Saturday only from 02.01.71.
		05.09.82	19.06.83	Sunday only allocation on daily route.
	J	13.06.70	03.09.71	Daily allocation.
	HT	04.09.71	24.10.86	Daily allocation. Monday to Saturday only from 04.09.82.
	FW	19.09.71	09.08.81	Sunday only allocation on daily route.
	S	26.06.83	21.04.85	Sunday only allocation on daily route.
28	X	08.12.63	03.03.89	Sunday allocation on daily route. Daily from 18.04.70. Monday to Saturday from 11.07.87
	WD	18.04.70	10.07.87	Daily allocation.
29	HT	07.09.68	13.12.75	Daily allocation. Monday to Saturday only from 24.01.70.
		19.03.77	04.11.88	Daily allocation. Monday to Saturday only from 07.02.87.
	WN	07.09.68	13.12.75	Daily allocation.
		19.03.77	04.11.88	Daily allocation. Monday to Saturday from 07.02.87.
	PB	08.09.68	18.01.70	Sunday only allocation on daily route.
	AD	24.03.69	04.11.88	Sunday only allocation on daily route. Daily from 19.03.77, Monday to Saturday from 23.04.83. Daily from 27.10.84, Monday to Saturday from 07.02.87.
30	AF	01.06.64	06.02.87	Daily allocation.
	H	01.07.64	24.04.81	Daily allocation.
	AG	25.04.81	03.09.82	Daily allocation.
	CT	04.09.82	06.02.87	Daily allocation.

Route	Garage	F.D.O.	L.D.O.	Notes
31	B	26.10.69	03.09.82	Sunday allocation on daily route. Daily from 17.06.72.
	CF	26.10.69	03.09.82	Sunday allocation on daily route. Saturday and Sunday from 07.02.70, daily from 17.06.72.
	X	04.09.82	14.04.89	Daily allocation. Monday to Saturday only from 21.11.87.
	HT	09.02.87	14.04.89	Monday to Friday only allocation on daily route.
32	W	13.06.70	12.03.71	Daily allocation.
	AE	17.01.71	07.03.71	Sunday only allocation on daily route.
33	U	02.09.65	30.12.66	*(Silvertown — Becontree)* Monday to Friday only route.
	M	31.12.66	24.06.83	*(Kensington — Hounslow)* Monday to Saturday only route. Daily from 04.09.82.
	FW	25.04.81	14.08.81	Monday to Saturday only route.
34	AD	17.01.71	04.09.77	Sunday allocation on daily route.
34B	AR	01.09.59	10.11.59	RMs 44, 45, 48, 49, 51, 56 trial service.
35	Q	07.09.68	20.06.86	Daily allocation. Saturday and Sunday only from 04.09.82, daily from 25.06.83. Monday to Saturday only from 03.08.85, daily from 02.11.85.
	H	07.09.68	24.04.81	Daily allocation.
	AG	25.04.81	20.06.86	Daily allocation.
	WL	04.08.85	26.10.85	Sunday only allocation on daily route.
36	PM	18.02.63	21.03.76	Daily allocation.
		12.01.80		Daily allocation. Monday to Saturday from 06.06.87.
36A	RL	15.01.63	21.03.69	Monday to Saturday only route.
	PM	22.03.69	26.03.76	Monday to Saturday only route. Monday to Friday only from 08.01.72.
		14.01.80	26.04.91	Monday to Friday only route.
36B	RL	15.01.63	21.03.69	Daily allocation.
	Q	22.03.69	02.01.72	Saturday and Sunday only allocation on daily route. Sunday only from 24.01.70.
	PM	22.03.69	12.04.76	Daily allocation.
		12.01.80		Daily allocation. Monday to Saturday only from 06.06.87.
	NX	24.01.70	01.01.72	Saturday only allocation on daily route.
	TL	02.11.85		Monday to Saturday only allocation on daily route.
37	SW	12.12.62	24.04.81	Daily allocation.
	AF	12.12.62	27.10.78	Daily allocation. Monday to Saturday only from 11.10.75.
	RL	23.12.63	16.03.69	Sunday only allocation on daily route.
	PM	23.03.69	02.01.72	Sunday only allocation on daily route. Daily from 17.04.71.
	NX	11.10.75	20.06.86	Daily allocation.
	AV	28.10.78	20.06.86	Daily allocation.
	CA	25.04.81	20.06.86	Daily allocation.
38	T	16.01.71		Daily allocation. Monday to Saturday from 06.06.87.
	CT	16.01.71	30.03.79	Monday to Saturday only allocation on daily route. Daily from 05.01.74, Monday to Saturday only from 11.10.75.
	CT	24.02.90		Monday to Saturday only allocation on daily route.
40	U	05.04.64	12.04.70	Sunday allocation on daily route.
	PR	27.01.65	06.09.68	Monday to Friday only allocation on daily route.
		26.10.68	04.07.84	Monday to Saturday only allocation on daily route. Saturday and Sunday only route from 18.04.70, daily from 22.04.78. Monday to Saturday allocation on daily route from 23.04.83.
	Q	31.01.65	02.08.75	Sunday only allocation on daily route. Saturday and Sunday only route from 18.04.70.
		23.04.78	24.04.81	Daily route.
	WH	18.06.66	25.10.68	Saturday only allocation on daily route. Saturday and Sunday only from 07.09.68.
40A	Q	27.01.65	21.04.78	Monday to Friday only route. Monday to Saturday only from 31.12.66, Monday to Friday only from 18.04.70.
	PR	27.01.65	21.04.78	Monday to Friday only route. Monday to Saturday only from 31.12.66, Monday to Friday only from 18.04.70.
40B	Q	30.01.65	24.12.66	Saturday only route.
	PR	30.01.65	24.12.66	Saturday only route.
41	WH	27.04.60	06.09.68	Monday to Saturday only allocation on daily route. (T/B 6).
	AR	13.10.63	01.11.85	Sunday allocation on daily route. Daily from 03.02.64. Monday to Friday only allocation on daily route from 04.09.82.
43	MH	20.06.63	25.01.75	Monday to Saturday only route.
		04.09.82	10.07.87	Monday to Saturday only route.
	FY	23.04.83	20.06.86	Monday to Saturday only route.
45	CF	10.11.63	07.01.72	Sunday allocation on daily route. Saturday and Sunday from 10.04.65, daily from 01.01.66. Monday to Saturday only allocation on daily route from 23.01.66.
	WL	01.12.65	22.02.75	Daily route. Monday to Saturday only from 31.12.66.
		21.03.81	02.08.85	Daily route. Monday to Friday allocation on daily route from 11.06.84. Saturday and Sunday only allocation on daily route from 27.10.84. Sunday allocation from 27.04.85.
	HT	25.04.81	02.08.85	Monday to Friday only allocation on daily route.
	SW	29.10.84	26.04.85	Monday to Friday only allocation on daily route.
46	AC	02.09.65	06.09.68	*(Alperton — Waterloo)* Monday to Saturday only route.
	X	09.09.68	24.10.69	Monday to Friday only route.
	CF	08.01.72	16.06.72	*(Hampstead Heath — Farringdon Street)* Monday to Saturday only route.
47	D	05.02.66	24.04.81	Saturday and Sunday allocation on daily route. Daily from 26.01.75.
	CT	16.01.71	01.01.72	Saturday allocation on daily route.
	TL	26.01.75	29.09.84	Daily allocation.
	TB	27.01.75	30.10.84	Monday to Friday only allocation on daily route. Daily from 04.09.82.
	AG	25.04.81	03.09.82	Daily allocation.
48	PR	11.11.59	26.01.65	*(Waterloo — N. Woolwich)* Daily allocation. (T/B 4)
	T	17.01.71	02.08.85	*(London Bridge — Whipps Cross)* Sunday allocation on daily route. Daily from 15.05.71.
49	AL	15.05.71	06.02.87	Monday to Saturday only allocation on daily route. Daily from 12.05.73, Monday to Saturday only from 28.10.78. Daily from 27.10.84.
	AK	15.05.71	26.10.84	Daily allocation.
		07.02.87	10.07.87	Daily allocation.
	S	29.04.85	10.07.87	Monday to Friday only allocation on daily route. Monday to Saturday only from 01.02.86, daily from 25.10.86. Monday to Saturday only from 07.02.87.

Route	Garage	F.D.O.	L.D.O.	Notes
51	SP	12.06.76	20.05.77	Daily allocation.
51A	SP	17.01.77	20.05.77	Monday to Saturday only route.
52	AC	03.10.65	24.10.86	Sunday allocation on daily route. Daily from 15.05.66. Monday to Saturday only from 15.08.81, daily from 04.09.82, Monday to Saturday from 02.11.85.
	GM	03.10.65	23.01.70	Sunday allocation on daily route. Daily from 12.06.66.
	X	24.01.70	17.04.83	Daily allocation. Sunday only allocation from 15.08.81.
52A	AC	23.01.66	18.12.66	*(Victoria — Borehamwood)* Sunday only route.
	X	15.08.81	03.09.82	*(Victoria — Westbourne Park)* Daily allocation.
	GM	04.09.82	15.08.86	Daily allocation. Monday to Saturday only route from 02.11.85.
53	AM	01.07.67	23.01.70	Daily allocation. Sunday to Friday only from 25.10.69.
	NX	01.09.67	08.01.77	Daily allocation.
		23.01.81	15.01.88	Daily allocation.
	PD	03.08.85	15.01.88	Daily allocation.
54	ED	11.03.73	15.04.78	Sunday allocation on daily route. Saturday allocation from 01.12.73.
	TL	22.03.75	15.04.78	Saturday allocation on daily route.
55	HL	03.10.65	29.11.68	*(Chiswick — Hayes)* Daily allocation.
	V	03.10.65	29.11.68	Daily allocation.
	T	17.01.71	27.10.72	*(Walthamstow — Marylebone)* Sunday allocation on daily route.
		31.01.81	05.06.87	Daily allocation. Monday to Saturday from 07.02.87.
56	PR	14.05.60	06.05.61	Saturday and Sunday only allocation on daily route. Saturday only from 15.10.60.
57A	BN	15.05.66	24.11.68	Sunday only route.
		20.07.69	27.12.70	Sunday only route.
58	WH	03.02.60	20.11.82	Daily allocation (T/B 5). Monday to Saturday from 25.04.81.
	WW	03.02.60	25.01.81	Daily allocation (T/B 5). Saturday and Sunday only from 16.06.73.
59	TC	30.08.64	23.11.75	Sunday only route.
	AK	14.06.70	22.10.78	Sunday only route.
59A	Q	04.01.71	10.03.72	Monday to Friday only route.
60	TC	04.09.82	22.04.83	Monday to Saturday only route.
62	BK	07.04.79	12.01.80	Daily allocation.
63	HT	01.02.61	23.01.70	Daily allocation (T/B 9). Monday to Friday only route from 09.10.63.
	PM	24.02.63	26.09.76	Sunday allocation on daily route. Daily from 17.07.63. Monday to Friday only route from 09.10.63. Daily from 24.01.70.
		04.09.82	02.11.85	Daily allocation.
63A	PM	13.10.63	18.01.70	Saturday and Sunday only route.
64	ED	20.07.60	08.05.62	Daily allocation (T/B 7).
	TH	09.05.62	03.12.71	Daily allocation. Monday to Saturday only from 29.01.64. Daily from 16.09.67.
	TC	29.01.64	05.11.67	Saturday and Sunday allocation on daily route. Daily from 01.09.64, Monday to Saturday only from 23.01.66.

Route	Garage	F.D.O.	L.D.O.	Notes
65	NB	19.10.75	03.09.82	Daily allocation.
		14.01.84	31.01.86	Daily allocation.
	K	20.10.75	13.01.84	Sunday only allocation on daily route. Daily from 31.03.79. Initially mainly vehicles loaned from NB.
	HL	04.08.85	26.01.86	Sunday only allocation on daily route.
66	NS	10.07.66	12.07.70	Sunday allocation on daily route.
66A	NS	10.07.66	13.07.69	Sunday allocation on daily route.
66B	NS	19.07.70	02.01.72	Sunday only route.
67	SF	19.07.61	03.12.71	Monday to Saturday only route (T/B 11). Between 10.07.66 & 24.01.70 was part allocation on Saturdays only)
68	CF	10.11.63	27.10.78	Sunday allocation on daily route. Saturday and Sunday from 31.12.66, daily from 07.02.70. Monday to Friday only allocation on daily route from 16.06.72.
		25.04.81	24.10.86	Monday to Saturday only allocation on daily route. Monday to Friday only from 27.10.84.
	TC	02.02.64	24.10.86	Sunday allocation on daily route. Saturday and Sunday from 31.12.66, daily from 07.02.70.
	N	01.11.64	04.01.74	Sunday allocation on daily route. Saturday and Sunday from 05.11.66, daily from 07.02.70. Monday to Saturday only allocation from 08.01.72.
	N	27.10.84	24.10.86	Daily allocation.
	WL	09.01.72	11.06.72	Sunday only allocation on daily route.
	Q	17.06.72	26.10.84	Saturday and Sunday only allocation on daily route. Daily from 05.01.74.
	TH	03.02.85	19.10.86	Sunday only allocation on daily route.
69	WH	03.02.60	30.07.84	Daily allocation (T/B 5). Monday to Saturday from 23.04.83.
	T	07.09.68	26.01.85	Monday to Saturday only allocation on daily route. Saturday only from 16.06.73, Saturday and Sunday only from 05.01.74. Saturday only from 11.10.75, Monday to Saturday only from 31.01.81. Saturday only from 04.09.82.
	PR	18.04.71	24.10.71	Sunday only allocation on daily route.
	WW	11.10.75	30.01.81	Saturday and Sunday only allocation on daily route. Daily from 31.01.76.
71	K	17.02.78	15.01.84	Monday to Saturday allocation on daily route.
	NB	14.01.84	02.08.85	Monday to Saturday allocation on daily route.
72	R	05.02.66	18.01.70	Saturday and Sunday allocation on daily route.
		21.10.73	30.01.81	Sunday allocation on daily route. Daily from 14.12. 75.
73	AV	05.12.62	01.10.65	Daily allocation. Sunday to Friday only from 08.05.63.
	AR	12.12.62		Daily allocation. Monday to Saturday only from 11.07.87.
	M	12.12.62	30.08.82	Daily allocation. Sunday only from 24.01.70.
	R	31.01.81	24.06.83	Daily allocation. Monday to Saturday only from 04.09.82
	S	25.06.83	12.08.88	Monday to Saturday only allocation on daily route. Daily from 11.07.87.
	MH	02.02.86	05.07.87	Sunday allocation on daily route.
74	HT	18.07.62	11.09.62	Monday to Friday only supplementary schedule. Baker Street Station — The Zoo.
		17.06.63	10.09.63	Monday to Friday only supplementary schedule. Baker Street Station — The Zoo.
		01.07.64	08.09.64	Monday to Friday only supplementary schedule. Baker Street Station — The Zoo.
		30.06.65	07.09.65	Monday to Friday only supplementary schedule. Baker Street Station — The Zoo.
		25.04.81	06.02.87	Daily allocation. Monday to Saturday from 03.08.85.
	AF	05.10.63	02.08.85	Saturday and Sunday allocation on daily route. Daily from 07.11.65.
	R	07.11.65	30.12.66	Saturday and Sunday only allocation on daily route.
74A	R	08.11.65	30.12.65	Monday to Friday only route.
74B	R	08.11.65	27.10.78	Monday to Friday only route.
75	TL	23.06.69	20.02.77	Sunday allocation on daily route.
76	AR	01.09.59	10.11.59	RMs 44, 45, 48, 49, 51, 56 trial service.
		09.02.64	01.02.85	Sunday only allocation on daily route. Daily from 01.11.65. Part allocation on Monday to Friday between 12.06.66 and 24.01.70. Includes FRM from 20.06.67 to 25.10.69. Monday to Saturday allocation on daily route from 29.01.83.
77	AL	17.12.73	31.01.86	Sunday allocation on daily route. Daily from 15.12.73, Monday to Saturday only from 27.10.84.
	SW	27.10.84	31.01.86	Saturday and Sunday only allocation on daily route. Monday to Friday allocation from 03.08.85.
77A	AL	17.12.73	26.10.84	Monday to Friday only route. Monday to Friday only allocation on daily route from 25.04.81.
	SW	17.12.73	02.02.85	Monday to Friday only route. Daily from 25.04.81.
77B	AL	03.10.71	06.05.73	Sunday allocation on Saturday and Sunday only route.
77C	AL	15.12.73	19.04.81	Sunday only allocation on Saturday and Sunday only route.
	SW	08.01.72	18.04.81	Saturday only allocation on Saturday and Sunday only route.
78	PM	21.07.63	07.05.72	Sunday allocation on daily route. Saturday and Sunday from 19.07.69.
81	AV	05.12.62	17.04.70	Sunday allocation on daily route. Saturday and Sunday from 21.11.64, daily from 23.08.69.
81B	AV	13.05.62	17.04.70	Sunday allocation on daily route (T/B 14). Daily from 12.06.63.
81C	AV	13.01.68	17.04.70	Saturday allocation on Monday to Saturday route. Monday to Saturday from 23.08.69.
83	ON	01.10.66	03.09.82	Daily allocation. Monday to Saturday from 15.07.71
85	AF	14.08.63	01.01.71	Monday to Saturday only route.
	NB	14.08.63	06.09.68	Monday to Saturday only route. Monday to Friday only allocation from 09.10.63.
		02.12.68	17.04.70	Monday to Friday only allocation on Monday to Saturday route.
	K	07.09.68	29.11.68	Monday to Friday only allocation on Monday to Saturday route.
85A	AF	14.08.63	09.03.73	Monday to Saturday only route.
86	U	05.04.64	12.04.70	Sunday allocation on daily route.
		28.10.72	28.08.82	Saturday only allocation on daily route.
	AP	28.02.76	24.07.84	Daily allocation.
	WH	23.04.83	26.01.85	Saturday only allocation on daily route.
87	NS	02.11.75	21.03.82	Sunday only allocation on daily route.
	BK	28.10.78	16.02.80	Daily allocation.
88	SW	12.06.66		Daily allocation. Monday to Saturday only from 25.04.81. Daily from 04.09.82. Monday to Saturday from 07.02.87.
	S	01.07.66		Daily allocation. Monday to Saturday only from 25.06.83, daily from 27.10.84. Monday to Saturday from 07.02.87.
	AL	26.04.81	26.01.86	Sunday only allocation on daily route.
89	NX	01.03.75	15.04.78	Saturday only allocation on daily route.
90	FW	18.07.70	30.01.72	Saturday and Sunday allocation on daily route. Sunday from 18.09.71.
90B	FW	19.07.90	05.01.73	Sunday allocation on daily route. Daily from 18.09.71.

Route	Garage	F.D.O.	L.D.O.	Notes
91	V	18.09.57	31.10.59	RM 2 trial service.
		12.08.59	10.11.59	RMs 36, 57, 75, 89, 94, 95, 96, 101, 111 trial service.
93	AF	06.06.64	11.04.70	Saturday allocation on Monday to Saturday allocation.
	AL	15.12.73	21.10.78	Saturday only allocation on daily route.
	A	28.03.76	03.09.82	Daily allocation.
94	TB	02.05.76	03.09.82	(Orpington — Brockley Rise) Saturday and Sunday allocation on daily route. Daily from 27.08.78.
	TL	22.04.78	03.09.82	Saturday allocation on daily route. Monday to Saturday only allocation on daily route from 27.08.78.
	S	22.09.90		(Acton Green — Oxford Circus) Monday to Saturday allocation on daily route.
95	BN	16.05.66	01.01.71	Monday to Saturday only route.
95A	BN	15.05.66	27.12.70	Sunday only route.
97	HL	10.05.61	26.09.65	(Ruislip — Brentford) Sunday only allocation on daily route. Saturday and Sunday allocation from 12.10.63. Sunday allocation from 18.11.64.
	SF	04.12.71	27.10.78	(Liverpool Street — Northumberland Park) Monday to Saturday only route.
99	AW	23.03.69	18.01.70	Sunday only allocation on daily route.
	AM	22.03.69	17.01.70	Saturday allocation on daily route.
100	U	01.04.64	08.10.71	Daily allocation. Monday to Friday only from 30.11.70.
101	U	30.10.71	28.04.78	Saturday and Sunday allocation on daily route. Daily from 28.10.72.
		28.10.79	03.09.82	Daily allocation.
102	MH	23.06.63	19.01.75	Sunday only allocation on daily route. Saturday and Sunday only from 24.01.70.
		22.02.75	22.10.78	Saturday and Sunday only allocation on daily route.
	AD	24.01.70	03.09.82	Saturday and Sunday allocation on daily route. Daily from 01.02.78.
104	FY	08.11.61	15.01.71	Daily allocation (T/B 12).
		24.03.73	21.10.78	Saturday only allocation on daily route.
	HT	12.11.61	03.09.82	Sunday only allocation on daily route (T/B 12). Saturday and Sunday only from 03.01.62, Saturday only from 09.05.62. Daily from 16.01.71. Monday to Friday allocation on daily route from 07.08.76. Daily from 25.04.81.
104A	FY	24.01.66	15.01.71	Monday to Friday only route.
105	HW	30.04.78	03.09.82	Daily allocation.
	S	01.05.78	30.01.81	Monday to Saturday only allocation on daily route.
106	CT	07.09.68	19.10.68	Saturday only allocation on Monday to Saturday route.
	H	07.09.68	11.08.72	Saturday allocation on Monday to Saturday route. Monday to Saturday from 02.12.68. Daily from 16.01.71.
		31.03.79	24.04.81	Daily allocation.
	PR	26.10.68	15.01.71	Saturday allocation on Monday to Saturday route. Monday to Saturday from 02.12.68.
	AR	16.01.71	11.08.72	Daily allocation.
		31.03.79	24.04.81	Daily allocation.
	AG	25.04.81	03.09.82	Daily allocation.
106A	D	19.05.63	10.01.71	Sunday only route.
	H	22.11.64	26.12.66	Sunday only route.
	H	08.09.68	10.01.71	Sunday only route.
108A	NX	22.03.69	18.01.70	Daily allocation.
109	BN	24.05.76	31.01.78	Monday to Friday allocation on daily route.
		15.08.81	06.02.87	Monday to Saturday only allocation on daily route.
	TH	04.10.76	27.08.78	Monday to Saturday allocation on daily route.
	TH	04.09.82	06.02.87	Daily allocation.
110	AV	10.07.66	18.12.66	Sunday allocation on daily route.
110A	AV	01.01.67	17.08.69	Sunday only route.
112	SE	14.06.70	09.05.71	Sunday allocation on daily route.
113	AE	23.12.62	24.10.86	Sunday allocation on daily route. Daily from 01.09.66.
115A	TH	06.12.64	27.12.70	Sunday only route.
116	AV	16.06.63	18.12.66	Sunday allocation on daily route.
		24.08.69	12.09.71	Sunday allocation on daily route.
116A	AV	01.01.67	17.08.69	Sunday only route.
117	AV	09.05.62	27.01.78	Daily allocation. (T/B 14)
	V	24.01.70	27.01.78	Monday to Saturday only allocation on daily route.
118	AK	17.06.72	27.10.78	Saturday allocation on daily route. Monday to Saturday only allocation on daily route from 14.12.75.
	BN	14.12.75	26.04.85	Daily allocation.
119	TB	03.05.76	30.10.84	Monday to Friday only route. Monday to Saturday only from 28.10.78.
119A	TB	01.02.75	21.10.78	Saturday only route.
119B	TB	26.01.75	14.10.84	Sunday only route.
122	AM	22.04.78	23.02.80	Daily allocation.
123	WW	27.04.60	18.03.77	Daily allocation. (T/B 6)
	AR	12.10.63	02.10.65	Saturday allocation on daily route. Monday to Saturday only allocation on daily route from 03.02.64.
	WN	07.09.68	13.12.75	Saturday and Sunday only allocation on daily route. Saturday only from 05.01.74.
	AD	24.03.69	18.03.77	Monday to Friday only allocation on daily route.
124	TL	24.07.71	07.01.72	Monday to Saturday only route.
124A	TL	25.07.71	02.01.72	Sunday only route.
127	EM	26.04.61	23.01.70	Monday to Saturday only allocation on daily route. (T/B 10)
	HT	26.04.61	18.01.70	Daily allocation (T/B 10). Sunday only allocation from 07.09.68.
	E	07.09.68	23.01.70	Daily allocation.
130	TC	01.09.64	01.03.75	Daily allocation. Monday to Saturday only allocation on daily route from 31.10.70. (Includes 130 Express)
	TH	01.11.70	28.11.71	Sunday only allocation on daily route.
		03.10.76	20.08.78	Sunday only allocation on daily route.
130A	TC	05.09.64	01.03.75	Saturday and Sunday only route. Saturday only allocation on Saturday and Sunday route from 23.01.66. Saturday only route from 16.09.67. Saturday only allocation on Saturday and Sunday route from 18.04.70. Saturday and Sunday allocation from 12.05.73. Saturday only from 05.01.74.
	TH	19.04.70	28.11.71	Sunday only allocation on Saturday and Sunday route.
130B	TC	01.09.64	28.02.75	Monday to Friday only route. Derivative RMs continued until 03.09.82.
130C	TC	16.09.67	17.04.70	Monday to Saturday only allocation on daily route.
	TH	12.11.67	12.04.70	Sunday only allocation on daily route.

Route	Garage	F.D.O.	L.D.O.	Notes
131	NB	09.05.62	11.05.73	Daily allocation. (T/B 14). Monday to Saturday only route from 23.01.66.
	AL	24.01.70	11.05.73	Monday to Saturday only route.
133	BN	24.05.76	03.03.78	Derivative journeys only.
		15.08.81	11.11.84	Daily allocation. Derivative journeys continued until 01.11.85.
134	MH	23.06.63	14.12.73	Sunday allocation on daily route. Daily from 14.07.64.
		04.09.82	18.05.86	Daily allocation.
	J	27.07.64	18.01.70	Monday to Friday allocation on daily route. Sunday to Friday only allocation from 23.08.69, Sunday only from 25.10.69.
	PB	12.08.64	14.12.73	Daily allocation.
		04.09.82	22.04.83	Daily allocation.
134A	MH	23.06.63	17.08.69	Sunday only route.
	J	26.97.64	17.08.69	Sunday only route.
135	E	17.01.78	26.09.80	*(Carterhatch — Brimsdown, Lockfield Avenue)* Monday to Saturday allocation on daily route.
	HT	21.11.87	04.11.88	*(Archway — Marble Arch)* Monday to Saturday only route.
137	GM	21.10.64	30.08.82	Daily allocation. Saturday and Sunday only allocation from 25.04.81.
	N	01.11.64	24.04.81	Daily allocation.
	CA	25.04.81	06.02.87	Daily allocation.
	AK	07.02.87		Monday to Saturday allocation on daily route. Monday to Friday only from 21.11.87.
	BN	07.02.87		Monday to Saturday allocation on daily route.
137A	N	17.04.65	16.09.73	Seasonal route.
140	AE	18.06.72	17.04.83	Sunday only allocation on daily route.
	HD	15.07.78	22.04.83	Daily allocation.
141	WN	08.11.61	18.03.77	Daily allocation (T/B 12). Monday to Saturday allocation on daily route from 14.12.75.
		08.04.82	01.02.85	Daily allocation. Monday to Saturday allocation on daily route from 27.10.84.
	NX	08.11.61	18.03.77	Monday to Friday only allocation on daily route. (T/B 12). Sunday to Friday only from 25.10.69. Daily from 04.09.71. Monday to Saturday only from 08.01.72.
		08.04.82	23.08.84	Monday to Saturday only allocation on daily route.
141A	NX	11.11.61	28.08.71	Saturday and Sunday only route (T/B 12). Saturday only from 25.10.69.
143	HT	01.02.61	29.11.68	Monday to Saturday only route (T/B 9).
144	WW	01.05.60	12.10.60	Sunday only allocation on daily route.
		12.05.63	04.01.74	Sunday only allocation on daily route. Saturday and Sunday allocation on daily route from 18.11.64. Sunday allocation from 25.10.69.
	AR	22.12.62	05.10.63	Saturday only allocation on daily route.
	WN	12.10.63	04.01.74	Saturday and Sunday allocation on daily route.
146	TB	30.01.77	16.04.78	Sunday allocation on daily route.
147	U	03.10.65	24.07.71	Saturday allocation on Monday to Saturday route.
148	AP	29.02.76	17.07.77	Sunday allocation on daily route.
149	EM	19.07.61	01.02.74	Daily route (T/B 11).
		27.09.80	31.01.86	Daily allocation. RCL from 27.09.80 to 14.12.84.
	SF	23.07.61	01.02.74	Sunday only allocation on daily route (T/B 11). Daily from 11.10.61, Monday to Saturday only from 28.02.62. Daily from 03.10.65.
		10.08.80	06.02.87	Daily allocation. RCL from 10.08.80 to 14.12.84.
	E	01.02.86	06.02.87	Daily allocation.
	AR	01.02.86	31.01.87	Saturday only allocation on daily route.
150	AP	28.02.76	14.10.77	*(Lambourne End — Becontree)* Saturday and Sunday allocation on daily route.
	U	02.05.87	31.08.87	*(Aldgate — Victoria)* Sunday only route.

Route	Garage	F.D.O.	L.D.O.	Notes
151	NX	09.11.74	21.04.78	Saturday allocation on Monday to Saturday route. Monday to Saturday from 26.01.75.
155	NB	23.01.66	12.04.70	Sunday only allocation on daily route.
		07.02.71	06.05.73	Sunday only allocation on daily route.
	AL	07.02.71	12.05.73	Sunday allocation on daily route.
		15.12.73	24.10.86	Saturday allocation on Monday to Saturday only route. Monday to Saturday allocation from 09.01.77.
159	AK	13.06.70	26.10.84	Monday to Saturday only route. Daily from 28.10.78.
		07.02.87		Monday to Saturday only allocation on daily route.
	Q	15.06.70		Mondy to Friday only allocation on Monday to Saturday route. Monday to Saturday from 31.10.70, daily from 28.10.78. Sunday to Friday only from 04.09.82, daily from 27.10.84. Monday to Saturday from 11.07.87.
	BN	05.09.82	25.11.84	Sunday only allocation on daily route.
		31.03.85	05.07.87	Saturday and Sunday only allocation on daily route. Daily from 25.10.86, Sunday only from 07.02.87.
		02.02.91		Monday to Saturday allocation on daily route.
	CA	27.10.86	06.02.87	Monday to Friday only allocation on daily route.
159A	Q	13.06.70	24.10.70	Saturday only route.
161	AW	02.03.75	30.10.81	Sunday only allocation on daily route. Daily from 27.09.80.
	SP	02.03.75	25.04.84	Sunday allocation on daily route. Daily from 21.05.77.
161A	AW	21.05.77	26.09.80	Monday to Saturday only route.
162	WH	03.02.60	30.12.66	Daily allocation (T/B 5). Monday to Saturday only route from 11.10.61, Monday to Friday only from 27.01.65.
	U	02.01.67	29.10.71	Monday to Friday only route. Monday to Saturday only from 18.04.70.
162A	WH	15.10.61	24.12.66	Saturday only route. Saturday only route from 30.01.65.
	U	31.12.66	23.10.71	Saturday only route.
164	AL	17.12.73	27.10.78	Monday to Friday only allocation on daily route. Vehicles derived from other routes.
	A	28.03.76	30.03.79	Saturday and Sunday allocation on daily route. Daily from 16.01.77.
164A	A	28.03.76	30.03.79	Saturday and Sunday allocation on daily route. Daily from 16.01.77.
165	RD	10.07.66	15.06.73	Daily allocation.
168	SW	31.08.75	27.10.78	Monday to Friday only allocation on Monday to Saturday route. Derivative RMLs only.
		16.12.80	24.04.81	Monday to Friday only allocation on Monday to Saturday route.
	WD	31.08.75	24.04.81	Monday to Friday allocation on Monday to Saturday route. Derivative RMs only until 16.12.80.
168A	HT	27.01.75	24.04.81	Monday to Friday only route. Single derivative RM only.
169	BK	07.03.64	11.07.70	Saturday and Sunday allocation on daily route. Saturday allocation from 07.09.68.
169A	U	04.04.64	12.07.69	Saturday allocation on Monday to Saturday route.
		01.03.76	19.03.77	Monday to Friday only route.
171	AR	08.02.64	31.01.86	Saturday allocation on daily route. Monday to Saturday allocation on Monday to Saturday only route from 07.09.68.
	NX	07.09.68	15.08.86	Monday to Saturday only route. Daily from 08.01.72. Monday to Saturday allocation on daily route from 03.08.85.
171A	NX	08.09.68	02.01.72	Sunday only route.
172	HT	04.09.71	19.01.75	Saturday and Sunday allocation on daily route.
		25.04.81	03.09.82	Daily allocation.
	Q	04.09.71	29.12.73	Saturday allocation on daily route.
		03.06.82	02.08.85	Monday to Friday only allocation on daily route.
173	RL	22.12.62	16.03.69	Saturday and Sunday allocation on daily route. Saturday and Sunday only allocation on daily route from 27.02.63. Sunday only from 11.04.65.
	PM	10.04.65	23.01.70	Saturday allocation on daily route. Monday to Saturday only allocation from 18.11.67, daily from 22.03.69.
174	NS	10.07.66	24.03.82	Daily allocation. (Includes 174 Express)
	RD	18.07.70	12.03.71	Daily allocation.
175	NS	11.10.75	03.09.76	Monday to Saturday allocation on daily route, RMA class.
		19.03.77	29.03.82	Monday to Saturday allocation on daily route.
175A	NS	05.10.74	07.01.77	Monday to Saturday only route.
176	AC	01.03.76	27.10.78	Monday to Friday only route.
	WL	22.03.76	08.06.84	Monday to Friday only route.
176A	WL	22.03.76	03.09.82	Monday to Friday only route.
177	NX	22.03.69	02.01.72	Sunday allocation on daily route. Saturday and Sunday from 24.01.70.
	AW	22.03.69	02.01.72	Sunday allocation on daily route.
179	BK	22.11.64	01.09.68	Sunday allocation on daily route.
180	AW	02.12.68	16.02.81	Monday to Saturday only route. Daily from 22.04.78.
	TL	02.12.68	30.10.81	Monday to Saturday only route. Daily from 22.04.78.
		04.09.82	26.10.84	Monday to Saturday only allocation on daily route.
180A	AW	09.01.72	16.04.78	Sunday only route.
	NX	09.01.72	16.04.78	Sunday only route.
181	SW	13.10.63	16.01.66	Sunday allocation on daily route.
		19.07.69	19.12.70	Saturday allocation on Monday to Saturday route.
183	AE	04.09.66	29.12.74	Sunday allocation on daily route.
184	WL	19.07.70	10.10.71	Sunday allocation on daily route.
185	WL	23.01.66	18.12.66	Sunday allocation on daily route.
187	X	07.12.63	07.06.70	Saturday allocation on daily route. Saturday only allocation from 03.10.65, Saturday and Sunday only from 31.12.66.
		24.02.75	03.09.82	Monday to Friday only allocation on Monday to Saturday route. Monday to Saturday from 15.08.81.
	ON	14.03.64	06.06.70	Saturday allocation on daily route. Saturday and Sunday from 01.10.66.
		24.02.75	03.09.82	Monday to Friday only allocation on Monday to Saturday route.
	SE	13.06.70	08.08.81	Saturday only allocation on Monday to Saturday route.
188	PM	20.07.63	15.03.69	Saturday allocation on daily route.
	NX	02.02.64	15.11.64	Sunday only allocation on daily route.
190	TC	18.04.70	21.09.75	Saturday allocation on daily route. Monday to Saturday from 20.11.71.
		28.02.76	26.01.85	Monday to Saturday only route. Daily from 28.10.78, Monday to Saturday from 23.04.83.
191	AR	08.02.64	09.07.66	Saturday only allocation on Monday to Saturday route.
		07.09.68	09.01.71	Saturday only allocation on Monday to Saturday route.

Route	Garage	F.D.O.	L.D.O.	Notes
192	NX	05.10.74	21.04.78	Saturday allocation on daily route. Monday to Saturday only allocation on daily route from 02.05.76.
	TL	26.01.75	16.04.78	Sunday only allocation on daily route.
193	AP	28.02.76	03.09.82	Saturday allocation on Monday to Saturday route. Monday to Saturday allocation from 03.10.76.
194B	ED	10.03.73	24.11.73	Saturday allocation on Monday to Saturday route.
196	Q	27.03.71	04.01.74	Monday to Saturday only route.
	N	05.01.74	24.04.81	Daily route.
	SW	25.04.81	03.09.82	Daily route.
197	TC	06.09.64	21.03.71	Sunday allocation on daily route.
		20.11.71	29.12.73	Saturday allocation on Monday to Saturday route.
207	HL	09.11.60	27.02.76	Daily allocation (T/B 8).
		17.08.80	27.03.87	Daily allocation.
	UX	08.05.63	27.03.76	Monday to Saturday allocation on daily route. Daily from 03.10.65.
		20.10.80	27.03.87	Daily allocation.
	HW	01.12.68	14.05.71	Sunday only allocation on daily route.
207A	HL	09.11.60	29.11.68	Monday to Saturday only route (T/B 8).
	HW	30.11.68	14.05.71	Monday to Saturday only route.
208	TL	04.09.82	06.07.84	Daily route.
212	MH	23.06.63	04.05.69	*(Muswell Hill — Finsbury Park)* Saturday and Sunday allocation on daily route.
	AD	14.06.70	12.09.71	*(Palmers Green — Staples Corner)* Sunday allocation on daily route.
213	NB	23.01.66	16.03.69	Sunday only allocation on daily route.
213A	NB	23.03.69	12.04.70	Sunday only allocation on daily route.
214	HT	01.02.61	25.01.75	Daily allocation (T/B 9).
220	S	20.07.60	01.01.71	Daily allocation (T/B 7).
221	FY	08.11.61	23.03.73	Daily allocation (T/B 12).
	WN	11.11.61	30.12.61	Saturday only allocation on daily route (T/B 12).
		13.05.62	23.03.73	Sunday only allocation on daily route. Sunday to Friday from 02.01.67, daily from 07.09.68.
225	WH	04.09.82	17.04.83	Saturday and Sunday only allocation on daily route.
	AP	06.09.82	22.04.83	Monday to Friday only allocation on daily route.
226	AC	10.11.68	07.06.70	Sunday allocation on daily route.
228	SP	08.03.75	27.01.78	Saturday allocation on daily route. Monday to Saturday from 21.05.77.
229	SP	21.05.77	03.09.82	Daily allocation.
230	T	16.06.73	30.01.81	Monday to Saturday only route.
233	TC	22.12.69	26.02.71	Monday to Saturday only route. FRM 1
234	TC	02.02.64	24.12.66	Sunday allocation on daily route. Saturday and Sunday from 01.09.64
		18.04.70	19.01.73	Monday to Friday only route. FRM 1
234B	TC	31.10.70	14.01.73	Saturday allocation on Saturday and Sunday route. FRM 1
237	AV	28.01.78	06.02.87	Daily allocation.
	V	28.01.78	18.07.86	Saturday only allocation on daily route.
238	PR	11.11.59	26.01.65	Monday to Friday only route (T/B 4).
	WH	11.11.59	16.04.71	Monday to Friday only route (T/B 4). Daily from 27.01.65.
238A	WH	14.10.61	23.01.65	Saturday only route.

Route	Garage	F.D.O.	L.D.O.	Notes
239	HT	01.02.61	30.12.66	Daily allocation (T/B 9). Monday to Friday only from 09.10.63.
	CF	02.01.67	23.01.70	Monday to Friday only route.
240	AE	23.12.62	07.06.70	Sunday only allocation on daily route.
	W	18.07.70	10.01.71	Sunday allocation on daily route.
241	WH	07.09.68	15.06.73	Daily allocation. Monday to Saturday from 30.10.71.
243	SF	19.07.61	28.08.82	Monday to Saturday only route (T/B 11). Saturday only allocation on Monday to Saturday route from 25.04.81.
	WN	02.01.67	03.09.71	Monday to Friday only allocation on Monday to Saturday route.
	AR	04.09.71	27.10.78	Monday to Friday only allocation on Monday to Saturday route. Monday to Saturday from 10.03.73.
		31.01.81	02.08.85	Monday to Saturday only route.
243A	SF	23.07.61	19.04.81	Sunday only route (T/B 11).
	AR	29.10.78	27.07.85	Sunday only route.
244	MH	25.10.69	09.01.71	Saturday allocation on Monday to Saturday only route.
245	W	03.01.62	12.06.70	Daily allocation (T/B 13). Monday to Friday only route from 18.11.64.
245A	W	21.11.64	31.08.68	Saturday and Sunday only route. Saturday only from 23.01.66.
249	WW	27.04.60	06.09.68	Daily allocation (T/B 6). Sunday to Friday only from 13.10.62 to 30.01.65.
	WH	27.04.60	06.09.68	Daily allocation (T/B 6). Sunday to Friday only from 13.10.62 to 30.01.65.
249A	WH	27.04.60	01.09.68	Daily allocation (T/B 6). Sunday only route from 15.10.61.
249B	WW	10.10.62	26.01.65	Monday to Saturday only route.
	WH	10.10.62	26.01.65	Monday to Saturday only route.
253	HT	01.02.61	09.10.75	Daily allocation (T/B 9). Monday to Friday from 26.01.75.
		22.04.79	26.10.85	Sunday only allocation on daily route.
	EM	26.04.61	10.10.61	Daily allocation (T/B 10).
	SF	11.10.61	20.11.87	Daily allocation.
	CT	08.01.72	03.09.82	Monday to Saturday only allocation on daily route. Daily from 11.10.75.
	D	28.10.78	24.04.81	Monday to Saturday only allocation on daily route. Daily from 27.09.80.
	AG	25.04.81	20.11.87	Saturday and Sunday only allocation on daily route. Daily from 04.09.82.
253A	HT	12.05.63	06.10.63	Sunday only route.
	SF	12.05.63	06.10.63	Sunday only route.
253B	HT	05.07.64	15.11.64	Sunday only route.
	SF	05.07.64	15.11.64	Sunday only route.
255	HL	09.11.60	02.10.65	Daily allocation (T/B 8).
	R	04.10.65	16.06.72	Monday to Friday only route. Daily from 31.12.66, Monday to Saturday from 15.02.69.
256	WW	03.02.60	21.04.78	Monday to Saturday only route (T/B 5). Monday to Friday only from 29.01.64 to 31.12.66.
256A	WW	01.02.64	24.12.66	Saturday only route.
257	WW	03.02.60	06.09.68	Daily allocation (T/B 5). Sunday to Friday only from 01.02.64 to 31.12.66.
	CT	23.03.68	06.09.68	Monday to Saturday only allocation on daily route.
259	HT	26.04.61	23.01.70	Daily allocation (T/B 10). Monday to Saturday from 23.01.66.
	WN	26.04.61	02.01.62	Sunday to Friday only allocation on daily route (T/B 10).
	AR	03.01.62	09.03.73	Daily allocation. Monday to Saturday only route from 31.12.66.
	EM	06.12.71	09.03.73	Monday to Friday only allocation on Monday to Saturday route.

Route	Garage	F.D.O.	L.D.O.	Notes
260	W	06.03.57	19.08.58	*(Colindale — Surrey Docks)* RM 1 trial service.
	FY	03.01.62	12.06.70	*(Barnet — Hammersmith)* Daily allocation (T/B 13).
		17.01.71	22.10.78	Sunday only allocation on daily route.
	SE	03.01.62	14.08.81	Monday to Saturday only allocation on daily route (T/B 13). Monday to Friday only from 07.09.68, Sunday to Friday only from 28.10.78.
	AE	23.12.62	10.01.71	Sunday only allocation on daily route.
	AC	07.09.68	11.07.84	Saturday only allocation on daily route. Daily from 28.10.78, MOnday to Saturday from 13.03.84.
260A	SE	23.01.66	01.09.68	Sunday only route.
261	AD	22.03.69	17.01.70	*(Barnet — Arnos Grove)* Saturday allocation on Monday to Saturday only route.
		16.01.71	12.03.77	Saturday allocation on Monday to Saturday only route.
		04.02.78	21.04.78	Monday to Saturday only route.
	TB	04.09.82	22.04.83	*(Orpington — Brockley)* Daily route.
262	T	31.12.66	18.04.81	Daily allocation. Monday to Saturday from 28.10.72. Saturday allocation from 16.06.73.
	WH	07.09.68	21.04.78	Daily allocation. Monday to Saturday from 28.10.72.
		14.10.79	24.04.81	Monday to Saturday allocation on daily route.
263	FY	16.01.71	14.07.72	Daily allocation.
	HT	16.01.71	14.07.72	Saturday and Sunday only allocation on daily route.
266	SE	03.01.62	15.04.79	Daily allocation (T/B 13). Monday to Saturday only allocation from 23.01.66, Sunday only allocation from 07.09.68.
	W	03.01.62	01.02.85	Daily allocation (T/B 13). Monday to Saturday only from 13.06.70, daily from 15.08.81, Monday to Saturday from 12.02.84.
	AC	07.09.68	22.10.78	Monday to Saturday only allocation on daily route. Daily from 13.06.70, Sunday only from 17.6.72.
		16.08.81	30.08.82	Sunday only allocation on daily route.
	R	17.06.72	28.08.82	Monday to Saturday only allocation on daily route. Saturday only from 28.10.78.
267	FW	09.05.62	17.09.71	Daily allocation (T/B 14).
268	S	20.07.60	30.12.66	Daily allocation (T/B 7).
269	WN	26.04.61	06.09.68	*(Enfield — Tottenham Court Road)* Daily allocation (T/B 10).
	WG	26.04.61	07.11.61	Monday to Friday only allocation on daily route (T/B 10).
	T	16.06.73	30.01.76	*(Leyton — Chingford)* Monday to Friday only route.
	WW	16.06.73	30.01.76	Monday to Friday only route.
271	HT	20.07.60	31.12.65	Daily allocation (T/B 7).
		10.07.66	15.01.71	Daily allocation.
272	WH	03.02.60	13.06.69	Daily allocation (T/B 5).
275	WW	27.04.60	06.09.68	Daily allocation (T/B 6).
	AR	12.10.63	24.12.66	Saturday only allocation on daily route.
276	HT	26.04.61	13.08.63	Monday to Saturday only route (T/B 10). Monday to Friday only from 10.10.62.
278	WH	27.04.60	27.10.72	Daily allocation (T/B 6). Monday to Saturday only route from 03.01.62.
278A	WH	06.01.62	13.07.69	Saturday and Sunday only route. Sunday only from 27.01.65.
279	EM	26.04.61	31.01.86	Daily allocation (T/B 10). Monday to Saturday only route from 28.10.78.
	E	24.01.70	25.09.87	Monday to Saturday only allocation on daily route. Daily from 18.07.70. Monday to Saturday only route from 28.10.78.
	HT	25.01.70	12.07.70	Sunday only allocation on daily route.
279A	EM	26.04.61	03.12.71	*(Flamstead End — Tottenham Hale)* Monday to Friday only route (T/B 10).
		29.10.78	26.01.86	*(Waltham Cross — Liverpool Street)* Sunday only route.
	E	29.10.78	20.09.87	Sunday only route.
281	FW	09.05.62	14.08.81	Daily allocation (T/B 14). Monday to Saturday from 27.03.71.
282	NB	09.05.62	30.12.66	Daily allocation (T/B 14). Monday to Saturday only from 18.11.64.
283	NB	09.05.62	23.01.70	*(Tolworth — Kingston)* Monday to Saturday only route (T/B 14).
	EM	28.10.78	26.09.80	*(Lower Edmonton — Hammond Street)* Daily route. Officially DM but mainly operated by RMs.
284	PR	11.11.59	11.10.60	*(Barking — Trafalgar Square)* Night route, except Saturday night/Sunday morning. Renumbered to N84 from 12.10.60. (T/B 4).
	PB	13.10.73	29.09.76	*(Potters Bar local)* Monday to Saturday only route. FRM 1
285	FW	09.05.62	26.03.71	Daily allocation (T/B 14). Monday to Saturday only from 31.12.66.
	NB	23.01.66	26.03.71	Monday to Saturday only allocation on daily route. Daily from 31.12.66, Monday to Saturday from 19.07.69.
293	X	03.01.62	16.06.67	Monday to Friday only route (T/B 13).
295	S	19.06.67	16.06.72	Monday to Friday only route.
298	WN	24.01.70	15.01.71	Monday to Saturday only route.
	AD	24.01.70	26.09.80	Saturday only allocation on Monday to Saturday only route. Monday to Saturday from 13.06.70, daily from 19.03.77.
298A	AD	26.01.70	12.06.70	Monday to Friday only route.
		16.01.71	26.09.80	Monday to Saturday only route.
	WN	15.06.70	15.01.71	Monday to Friday only route.
299	WH	27.04.60	11.10.60	*(Chingford Mount — Victoria & Albert Docks)* Nightly allocation (T/B 6). Renumbered to N99 from 12.10.60.
	PB	16.01.71	27.06.71	*(Southgate — Borehamwood)* Saturday and Sunday allocation on daily route.
N68	WD	26.04.75	28.10.83	Sunday to Friday night only allocation on nightly route. Officially converted to DM from 25.10.75 to 15.12.80 but usually worked by RMs in practice.
	SW	26.04.75	28.10.83	Saturday night/Sunday morning only allocation on nightly route. Officially converted to DM from 25.10.75 to 15.12.80 but usually worked by RM/RML in practice.
N81	SW	02.01.71	28.10.83	Saturday night/Sunday morning only route. Usually RT worked until 07.01.72. Officially converted to DM from 25.10.75 to 15.12.80 but usually worked by RMs in practice.
N82	NX	15.06.68	02.11.79	Nightly allocation.
N83	SF	19.07.61	22.03.75	Nightly allocation (T/B 11). RMs continued to work the route unofficially until 11.83.
N84	PR	12.10.60	04.01.74	Sunday to Friday night only route (previously numbered 284).
N85	RL	17.12.62	21.03.69	Sunday to Friday night only route.
	PM	23.03.69	12.05.72	Sunday to Friday night only route.
N86	PM	18.02.63	12.05.72	Sunday to Friday night only route.
N87	BN	16.05.66	01.01.71	Sunday to Friday night only route.
	SW	13.06.66	**	Sunday to Friday night only route. ** Date not recorded, crew-DMS used by local arrangement sometime in 1982/83.

Route	Garage	F.D.O.	L.D.O.	Notes
N88	WD	19.04.70	24.10.75	Sunday to Friday night only route.
		16.12.80	03.09.82	Sunday to Friday night only route.
N89	R	01.02.66	03.09.82	Sunday to Friday night only route.
N90	AR	12.12.62	13.04.84	Sunday to Friday night only route. Nightly from 26/27.09.80.
N91	AC	03.10.65	29.03.81	Sunday to Friday night only route.
N92	J	17.10.63	03.09.71	Sunday to Friday night only route.
	HT	04.09.71	24.01.75	Sunday to Friday night only route.
		25.04.81	**	Sunday to Friday nightly only route. ** Date not recorded, crew-DMS used by local arrangement from sometime in 1982/83.
N93	HT	31.01.61	24.01.75	Sunday to Friday night only route (T/B 9).
		25.04.81	**	Sunday to Friday night only route. ** Date not recorded, crew-DMS used by local arrangement from sometime in 1982/83.
N94	W	24.12.62	01.03.74	Nightly allocation. Sunday to Friday night only from 31.01.71
N95	BK	01.07.64	17.07.70	Sunday to Friday night only route.
N96	T	31.12.66	18.03.77	Sunday to Friday night only route.
N97	V	25.01.70	19.09.75	Sunday to Friday night only route.
N98	AP	29.02.76	08.10.76	Nightly allocation.
N99	WH	12.10.60	03.09.82	Nightly allocation (previously numbered 299). Unofficially converted to DM from 09.07.78 but RMLs normally used in practice.
109	TH	**	**	Dates not recorded
168 Night	SW	19.04.70	25.04.75	Saturday night/Sunday morning only allocation on nightly route. Renumbered N68.
	WD	19.04.70	25.04.75	Sunday night to Friday night only allocation on nightly route. Renumbered N68.
177 Night	NX	**	**	Dates not recorded
181 Night	SW	**	**	Dates not recorded.
500	GM	26.11.84	20.12.84	Occasional RMs augmented the service during closure of the Victoria Line.
609	HT	30.04.61	12.11.61	Sunday only allocation on daily trolleybus route (T/B 10).
K1	K	12.11.83	12.01.84	Monday/Thursday/Saturday only route.
	NB	14.01.84	31.01.85	Monday/Thursday/Saturday only route.
K2	K	12.11.83	13.01.84	Tuesday/Friday/Saturday only route.
	NB	14.01.84	01.02.85	Tuesday/Friday/Saturday only route.
—	SW	07.04.79	28.09.79	(Shop-Linker) Monday to Saturday only route.
—	NX	16.04.62	28.10.62	(Circular Tour of London) Daily allocation.
		14.04.63	08.10.63	Daily allocation.
—	SW	31.01.78	03.02.83	(Round London Sightseeing Tour] Sporadic operation by FRM 1
—	B	22.03.86	16.04.88	(Official London Sightseeing Tour)
	WD	17.04.88		

APPENDIX 5
RMs RE-REGISTERED BY LONDON BUSES, 1985-1989

Between 1985 and 1989 a number of RMs were re-registered whilst in London Buses' ownership though not all have operated with their new registrations. They were:

RM				RM			
9	New reg OYM374A	Old reg to L 263		77	New reg OYM503A	Old reg to MT 6
12	New reg OYM413A	Old reg to M 1437		88	New reg OYM432A	Old reg to M 1379
14	New reg OYM424A	Old reg to L 262		100	New reg ALA814A	Old reg to privately-owned coach
15	New reg KGH602A	Old reg to M 1315		136	New reg OYM583A	Old reg to M 1436
20	New reg OYM378A	Old reg to L 260		400	New reg OYM518A	Old reg to DA 1
29	New reg OYM453A	Old reg to L 95		434	New reg OYM579A	Old reg to M 1434
31	New reg OYM611A	Old reg to MA 101		456	New reg XMC223A	Old reg to privately-owned car
46	New reg OYM580A	Old reg to M 1046		1002	New reg OYM368A	Old reg to L 261
53	New reg OYM582A	Old reg to M 853		2001	New reg EYY128B	Old reg to T 1000

APPENDIX 6
ALLOCATION OF ROUTEMASTERS ON FIRST DAY OF NEW OPERATING COMPANIES, 1st APRIL 1989

CENTREWEST

Routes operated		Mon-Fri		Sat
7	Westbourne Park	12 RML	Westbourne Park	10 RML
15	Westbourne Park	12 RML (+ 1 ex rte 7)	—	
31	Westbourne Park	11 RM	Westbourne Park	14 RM

Buses licensed for service
Westbourne Park   22 RM, 25 RML
Licensed trainers   4 RMC

EAST LONDON

8	Bow	37 RML	Bow	33 RML
15/B	Upton Park	47 RML	Upton Park	34 RML
X15	Upton Park	6 RMC	—	

Buses licensed for service
Bow   4 RM, 39 RML
Upton Park   3 RM, 52 RML, 6 RMC, 1 RMC (open top, special duties)
Licensed trainers   5 RM, 6 RMA, 2 RMC, 1 RML

LEASIDE BUSES

73	Tottenham	35 RML	Tottenham	28 RML

Buses licensed for service
Palmers Green   1 RM (special duties)
Tottenham   6 RM, 37 RML
Licensed trainers   8 RM, 1 RMA, 1 RML

LONDON CENTRAL

3	Camberwell	15 RM	Camberwell	13 RM
12	Camberwell	15 RML (+ 1 ex rte 159)	Camberwell	9 RML
	Peckham	{ 22 RML	Peckham	22 RML
		5 RM (+ 2 ex rte 36B)		
36	Peckham	25 RM	Peckham	19 RM
36A	Peckham	7 RM	—	
36B	Peckham	5 RM	Peckham	11 RM
159	Camberwell	{ 16 RML	Camberwell	14 RML
		10 RM		

Buses licensed for service
Camberwell   29 RM, 31 RML
Peckham   54 RM, 22 RML
Licensed trainers   9 RM

LONDON FOREST

6	Ash Grove	25 RML	Ash Grove	15 RML
38	Leyton	40 RML	Leyton	30 RML

Buses licensed for service
Ash Grove   1 RM, 28 RML
Leyton   1 RM, 45 RML
Licensed trainers   4 RM

LONDON GENERAL

9	Victoria	12 RM	Victoria	8 RM
10	Victoria	2 RM (+ 4 ex rte 19)	—	
		& 1 ex rte 11)		
11	Victoria	29 RM (+ 1 ex rte 10)	Victoria	20 RM
14	Putney	25 RML (+ 3 ex rte 22)	Putney	17 RML
19	Victoria	13 RM	Victoria	11 RM
22	Putney	18 RML	Putney	11 RML
88	Stockwell	13 RML	Stockwell	13 RML

Buses licensed for service
Putney   48 RML
Stockwell   15 RML
Victoria   69 RM
Licensed trainers   7 RM, 4 RMC

LONDON NORTHERN

Routes operated		Mon-Fri		Sat
18	Finchley	22 RML	Finchley	20 RML
19	Holloway	15 RML (+ 3 ex rte 31)	Holloway	8 RML
31	Holloway	7 RM		

Buses licensed for service
Finchley	24 RML		
Holloway	10 RM, 15 RML		
Licensed trainers	4 RMC		

LONDON UNITED

		Mon-Fri		Sat
9	Stamford Brook	⎰ 16 RML ⎱ 2 RM	Stamford Brook	12 RML
10	Shepherds Bush	14 RML	Shepherds Bush	21 RML
12	Shepherds Bush	6 RML	Shepherds Bush	5 RML
88	Shepherds Bush	15 RML	Shepherds Bush	10 RML

Buses licensed for service
Shepherds Bush	39 RML
Stamford Brook	5 RM, 16 RML
Licensed trainers	8 RM, 7 RMC

METROLINE

		Mon-Fri		Sat
6	Willesden	9 RML	Willesden	9 RML

Buses licensed for service
Willesden	1 RM, 11 RML
Licensed trainers	4 RM, 3 RMC, 1 RML

SELKENT

		Mon-Fri		Sat
36B	Catford	18 RM	Catford	15 RM

Buses licensed for service
Catford	21 RM
Licensed trainers	7 RM

SOUTH LONDON

		Mon-Fri		Sat
2B	Norwood	⎰ 17 RML ⎱ 1 RM	Norwood	16 RML
137	Streatham	7 RML	—	
	Brixton	25 RML	Brixton	20 RML
159	Streatham	18 RM	Streatham	10 RM

Buses licensed for service
Brixton	28 RML
Norwood	3 RM, 17 RML
Streatham	23 RM, 7 RML
Licensed trainers	9 RM, 2 RMC

CORRECTIONS TO VOLUME ONE

CHAPTER 12
The interior view of an RML reproduced on page 82 of Volume One should have been this view of RML 880. (London Transport Museum photo H/16675).

APPENDIX 4
The initial allocations for AF garage in December 1962 should read RM 1211, 1213-1225. Those for RL garage in January 1963 should read RM 1383-1407, 1409-1411.

INDEX

A

Abbey District	30, 105
Abbey National	70
Abbey Wood	21, 30
Active Ride Control (ARC)	60, 61
Acton	90, 103
Adams Foods	46
Addis	70
Addlestone	35, 36, 44
AEC	14, 15, 19, 20, 35, 41, 45, 57, 84, 85, 102, 108, 114, 117, 122
Ailsa Volvo	61, 98
Aldenham	12, 20, 29, 31, 32, 35, 36, 37, 38, 40, 48, 49, 50, 68, 70, 77, 79, 81, 82, 85, 88, 89, 90, 92, 95, 96, 97, 108, 112, 116, 117, 118, 120, 121, 126
Aldwych	53, 106, 123
Alexander	61, 110, 131
Allmey, Don	102
Alperton	17, 32
Amersham	34, 43
Amey Roadstone	70
Amsterdam	123
Andover	109
Apsley Mills	41
Ash Grove	30, 31, 32, 55, 102, 103, 105, 106
Ashfield, Lord	10
Atlantean	10, 11, 12, 36, 37, 43, 44, 45, 46, 47, 48, 62, 64
Austrian Wines	52
Automotive Products Ltd	60, 61
Avia Watches	70

B

BEA	27, 66, 67, 92
BEL (Bus Engineering Ltd)	90, 96, 108, 118
BOAC	66
BUSCO	27, 33, 102, 104
Bank of Cyprus	52
Barker & Dobson	52, 53
Barking	11, 19, 24, 25, 26, 27
Barnet	32
Barnsley	82, 84, 121
Barton	45
Battersea	14, 26, 27, 47, 55, 71, 74, 76, 83, 100, 102, 118, 119
Beckton	91, 111
Bedford	58, 127
Beijing	126
Bensham	63
Berkshire	11
Berry, Vic	121
Bertorelli Ice Cream	52, 55
Betjeman, Sir John	74, 75
Bexleybus	108
Bexleyheath	23, 103, 119
Birds	121
Blackpool	125, 128
Blue Arrow	34
Blue Bus Service	119, 125
Blue Triangle	116
Boeing	66
Booth	121
Boulogne Chamber of Trade	52
Bournemouth	41, 116
Bow	14, 18, 25, 108
Bradford-upon-Avon	58
Brakell, E.H.	85, 116
Brighton	124, 125
Bristol	41
Bristol LH	21, 80, 122
Bristol VR	43
British Airways	18, 53, 66, 67, 80, 81, 82, 84, 85, 121
British Leyland Unipart	52

British Motor Corporation	58
Brixton	12, 20, 23, 30, 31, 101, 102, 105
Bromley	16, 17, 20, 23, 24, 29, 31, 32, 98, 100, 101
Buchanan Street	131
Buckley, C.R.	34
Bulmers	70
Burberry	77
Burnley	121, 125, 127
Bus Electronic Scanning Indicator (BESI)	18
Busplan 78	24, 25, 26

C

CAV	16, 31
Camberwell	11, 12, 14, 31, 33, 77, 103, 105, 109, 112
Camden Town	76, 102
Canada	121
Capital Radio	104
Cardinal	32, 91, 109
Carlisle	129
Carlton	119
Carlyle	119
Catford	10, 12, 14, 16, 17, 23, 24, 28, 31, 32, 33, 98, 100, 102, 113
Celebrity Travel Agents	52
Central Bus Department	12
Centrewest	111
Certificate of Fitness	16, 23, 26, 27, 35, 38, 40, 44, 45, 64, 83, 92, 121
Chadwell Heath	24, 26
Chalk Farm	10, 14, 18, 78, 79, 102, 103, 110
Charles, Prince	29, 77
Chase Cross	81
Chelsham	34, 42, 43, 44, 45, 46
Cheshunt	58
China	126
Chingford	14, 19
Chipping Ongar	16
Chislehurst	22
Chiswick	15, 27, 28, 29, 40, 48, 49, 57, 58, 61, 66, 68, 69, 70, 71, 79, 81, 82, 85, 86, 87, 90, 95, 96, 103, 108, 118, 121
Citilink	125
Citybus	61, 125, 126
Clapham	25, 30, 78, 98, 99, 100, 103, 105
Clapton	12, 87, 105, 106
Clayton-Dewandre	26, 31
Clipper (East Yorkshire)	128
Clydeside	110, 124, 125, 130, 131
Cobham Museum	47, 68
Cockfosters	38
Colt 45 Lager	100
Commutabus	58
Corby	127, 129
Costers	125
County Hall	10
Covent Garden	49, 68, 69, 89, 102, 122
Cox, D.A.	62
Crawley	34, 36, 37, 41, 43, 44, 47
Crich	129
Cricklewood	11, 12, 15, 16, 27, 29, 33, 86, 100, 106
Crosville	34
Crown	70, 71
Crowthorne	11
Croydon	10, 11, 12, 18, 19, 28, 29, 31, 32, 47, 48, 78, 95, 98, 99, 100, 103
Cumberland	125, 129

D

DAF	108, 109, 119
Daily Mirror	70
Daimler	37, 41, 122
Dalston	16, 17, 18, 30, 68
Danone Yoghurt	52
Dartford	34, 36, 37, 39, 41, 42, 43, 44, 119

Denton	43
Deptford	17
Deregulation Day	124, 125
Diana, Lady	29, 77
Diddlers	114
Docklands	14
Dolphin International Displays Ltd	120
Dominic, Peter	52
Dorking	34, 36, 43, 44, 47
Double Deck Tours Ltd	123
Dundee	70, 124, 125, 131
Dunton Green	35, 36

E

EEC	92
Earl's Court	58
Eastenders	127
East Grinstead	40, 43, 47
East Kilbride	59
East Midland	125, 129
East Surrey Traction Company	34
East Yorkshire	125, 128
Eastbourne	41, 42
Easterhouse	124
Eastern Coach Works (ECW)	97
Eastern National	36
Eccles	119
Edgware	24
Edmonton	10, 14, 16, 25, 27, 49, 88, 89, 90, 99, 100, 103
Elizabeth II	70
Elm Park	111
Elmers End	14, 103
Endless Holdings	47
Enfield	10, 23, 25, 27, 103, 105
Ensign Bus Co	32, 85
Epsom	36
Esso	52, 74
Esso Uniflow	52
Evening News	52
Everton Mints	53
Executive	14, 16, 25, 56, 80, 82
Exide Batteries	70
Express	21, 45, 46
Essex County Council	21

F

Farley's Rusks	70
Federal Mogul	16, 80
Feltham	103
Fiat	109
Finchley	11, 14, 19, 25, 27, 32, 85, 102, 110, 130
Fine Fare	39
Firth, Donald	59
Fleetline	10, 11, 12, 15, 17, 19, 22, 24, 25, 27, 28, 29, 30, 37, 38, 41, 48, 56, 59, 82, 98, 106, 109, 122
Forest District	29, 90, 101, 106
Forest Ranger	78, 91
Frontsource Ltd	90
Fulham	120
Fulwell	12, 14, 30, 32, 85, 86, 100
Fylde	128

G

Gagg, M.W.	125
Gala Cosmetics	80, 121
Gangzhou	126
Garston	37, 38, 39, 40, 41, 43, 44, 45, 47
Gascoigne Estate	26
Gash, W. & Sons Ltd	129
Gatwick	108
Gibson	71
Glasgow	109, 124, 125, 130, 131

GLC	10, 15, 20, 21, 23, 25, 27, 29, 31, 32, 33, 50, 57, 59, 74, 92, 99, 103
Gloucester Road	66, 67, 84
Goddard's	70
Godstone	35, 36, 37, 38, 40, 41, 42, 44, 45, 47, 115
Gold Arrow	111
Golders Green	30
Grays	35, 36, 37, 39, 41, 42, 44, 45, 85, 86
Green Line	34, 35, 36, 38, 41, 42, 43, 44, 45, 46, 89, 90, 99, 118, 119, 122
Grey-Green	110
Grove Park	104
Guildford	34, 36
Guildhall	74, 75

H

HMV	77
Hackney	14, 26, 30, 51, 106
Hampshire Bus	129
Hanimex	52
Hanwell	19, 27, 31, 105, 121
Harlow	19, 36, 37, 41, 43, 44, 45
Harriman, Sir George	58
Harrow Weald	24, 28, 32, 106
Hatfield	35, 36, 38, 39, 40, 41, 43, 44, 45, 46
Havering	25, 61
Heathrow	66, 84
Heinz	70, 72
Hemel Hempstead	36, 41, 42, 43, 44
Hendon	19, 83, 92, 103, 105
Hertford	35, 36, 37, 41, 42, 43, 44, 46, 89
Hertfordshire County Council	42
High Court	32
High Wycombe	35, 37, 39, 43, 47
Highgate	11, 12, 29, 51, 68, 120
Holborn Circus	29
Holland	108
Holloway	11, 12, 14, 17, 18, 21, 22, 23, 29, 30, 31, 32, 89, 98, 100, 101, 103, 104, 105, 106, 107, 109, 110
Homepride Flour	52
Homerton Coaches Ltd	58
Hong Kong	126
Horlock	119
Hornchurch	14, 110
Hounslow	11, 23, 25, 103, 105, 122
House of Commons	31
House of Lords	29
Hove	125
Hull	125, 128
Hyde Park	18, 77, 79, 94, 101, 115
Hyde, Walter	50
Hydrolastic	58
Hyrostatic	59
Hyper-Hyper	123

I

ICL	70
Industrial & Machine Diesels	109
Interflora	120
International	57
Invicta Co-op	39
Iveco	109

J

JVC	70
Jarrow	63
Johnston	124
Jubilee	74, 79

K

Kelvin	125, 129
Kensington	123
Kent	108, 118, 119
Kingston	18, 23, 32, 33, 86, 100
Kleenex	70
Kosset	70

L

LBL	109, 118, 119
LCBS	23, 82, 84, 86
LGOC	103
LPC Coachworks	122
LPG (Liquid Petroleum Gas)	59
LRT	33, 98, 100, 105, 106, 108, 110
Ladbroke Grove	72
Ladbrokes	51, 52, 55
Lambert & Butler	70, 72
Leam Lane Estate	64
Leamington Spa	60, 68
Leaside	27, 104
Leatherhead	34, 39, 41, 44, 45
Leaver, J.	118, 119
Lee Valley Regional Park	106
Leeds	30, 76
Lewisham	113
Lex Tillotson	121
Leyland	16, 18, 19, 23, 26, 27, 31, 32, 41, 42, 43, 45, 49, 56, 57, 58, 59, 62, 74, 80, 97, 103, 104, 112, 117, 121, 125, 128
Leyton	12, 14, 19, 21, 29, 30, 82, 91, 105, 111,113
Lincoln Green	37, 38, 45, 47
Lincolnshire Road Car	129
Liverpool Street	25, 50
Lockheed	26, 31, 60, 68, 80
Locomotors	108, 109
Lodekka	62, 122
London Bridge	12
London Buses	64, 95, 96, 100, 101, 103, 104, 107, 108, 109, 110, 112, 116, 117, 119
London Bus Preservation Group	68, 69
London Coaches	79, 118, 119
London Country	11, 14, 19, 26, 27, 29, 34, 35, 37, 38, 39, 40, 43, 44, 46, 47, 49, 68, 74, 80, 82, 83, 84, 86, 121
London Country Bus (South West) Ltd	47
London General	114
London & Manchester Assurance	39
London Transport Museum	49, 68, 102, 122
London United	114, 115
Londoner	11
Long Hill	128
Loughton	103
Lowestoft	97
Luton	34, 43
Lyle & Scott	53, 55

M

MCW	16, 18
Magicbus	124, 125
Magnet (Clapton) Ltd	50
Maidstone	41, 42, 43, 125
Manchester	11, 119, 122, 125, 131
Mancunian	11
Manor House	14, 15, 87
Mansfield & District	129
Marble Arch	74, 90, 106, 107, 118
Masons Hill	31
Matlock	129
Meccano (Dinky Toys)	52
Mercedes	59, 110, 111, 113
Merlin	10, 15, 16, 29, 35, 41, 49, 56
Merton	12, 15, 21, 103, 105
Metrobus	23, 25, 27, 30, 32, 33, 59, 61, 98, 99, 100, 101, 102, 103, 109, 114, 116
Metroline	111
Metropolitan	16, 19, 21, 27, 30, 32, 80, 118
Metropolitan Transport Authority	31
Middle Row	11, 16, 17, 30, 77, 120
Midland Bank	33
Midland Red	116
Mini	58
Monkwearmouth	63
Moorgate	114
Morden	15
Mortlake	28, 32, 60, 61, 83, 122
Motability	89
Moulton, Dr. Alex	58
Multi-Ride	25
Museum of British Transport	30, 122
Muswell Hill	12, 15, 17, 18, 25, 31, 32, 93, 102, 103, 105
Myson	52, 53

N

NBA (No Bus Available)	14, 24, 26, 83
NGT	121
Napsbury Hospital	43
National	23, 31, 40, 41, 42, 43, 44, 45, 49
National Bus Company (NBC)	34, 38, 39, 41, 42, 43, 44, 47, 62, 64, 86, 97, 122, 125, 128, 129
National Engineering Laboratory (NEL)	59
NatWest Bank	70, 71
Negombo	130
Nescafé	70, 73
New Ash Green	119
New Bus Grant	56
New Cross	13, 17, 18, 20, 21, 24, 27, 29, 31, 71, 98, 102, 103, 108
New Street Square	38
Newcastle	64
Newark	129
Niagara Falls	123
Norbiton	10, 14, 18, 19, 32, 33, 102, 108
Northfleet	37, 38, 39, 40, 41, 42, 43, 44, 45, 47
North Street	16, 18, 19, 21, 31, 80, 81, 86
North Thames Gas	74, 76
North, W.	30, 32, 121
Northern	62, 64, 80, 83, 85, 116, 118, 121, 125
Northern Counties	37
Northern Rose	63
Norwood	10, 30, 72, 98, 99, 100, 103, 105, 109
Nottingham	97, 125
Nunhead	116

O

OLTST	117, 118
Obsolete Fleet	85, 116, 118
Olympian	97, 104, 118
Omnibus	74, 76
Orwell, George	33
Oxford Bus Museum	68
Oxford Circus	91, 98, 109
Oxford Street	75, 90

P

PVS (Barnsley) Ltd	121
Paisley	124
Palmers Green	10, 12, 21, 23, 24, 27, 30, 32, 61, 103, 104, 110, 113
'Parcel' Bus	77
Park Royal	35
Parliament	32
Pears	53
Peckham	13, 14, 19, 20, 21, 23, 27, 29, 32, 33, 57, 102, 104, 105, 109, 120
Perkins	58
Perth	124, 125, 131
Philip, Prince	70
Piccadilly	66, 131
Pimlico	10, 79
Plaxton	58
Plumstead	10, 24, 27, 30, 31, 32, 85, 89, 102, 108

Polo	36
Ponders End	23
Poplar	12, 26, 30, 33, 100, 102
Port of London Authority	84
Portsmouth	125
Potters Bar	15, 25, 27, 31, 32, 48, 49, 106, 113
Provincial	125
Purfleet	32
Putney	14, 104, 105, 106
Pye	52

Q

Quant, Mary	121

R

Radlett	46
Rambo	124
Rand	52, 53
Red Admiral	125
Red Arrow	10, 98, 110
Red Bus Rover	11
Redhill	39
Reed, Austin	77
Rees	118
Regent	41, 42, 77, 90
Reigate	34, 36, 39, 40, 43, 44, 45, 47, 108
Reliance	35, 45, 102
Ribble	43
Ridley, Nicholas	33
Riverside	14, 18, 24, 25, 29, 32, 33, 34, 36, 50, 51, 79
Roadliner	41
Rolls-Royce	62
Romford	34, 35, 36, 40, 43, 80
Rotherham	121
Royal Albert Dock	84, 88
Royal Blue	41
Royal College of Art	50
Rumplestiltskin	124

S

SCG	31
Sampsons Coaches & Travel Ltd	58
St Albans	34, 36, 39, 43, 90, 120
St Annes	128
St Enoch	124
St Katherine's Dock	70
St Mary Cray	103
St Paul's	70, 71
Scandinavia	121
Scania	16
Scotch House	76
Scottish Bus Group	95, 97, 109, 124, 125, 131
Scottish & Newcastle Breweries	51
Selfridges	70, 75
Selkent	111
Seven Kings	19, 20, 31, 32, 33, 78, 105
Seven Sisters	90
Sharp Electronics	52
Shenzhen	126
Shepherds Bush	11, 12, 14, 17, 24, 29, 32, 51, 99, 102, 103, 105, 110, 114, 115
Shillibeer, George	27, 71, 74, 75, 76, 83, 89, 96
Shirebrook	129
Shoplinker	27, 74, 75, 76
Sidcup	18, 20, 21, 22, 23, 28, 31, 32, 33, 78, 100, 102, 108
Sightseeing Tour	27, 48, 49, 85, 95, 116
Silexine Paints	50, 51, 52
Silver Jubilee	23, 70, 71, 73, 74
Simms	16, 31, 121
Sinclair-Cunningham	59
Singer	70
South Wales	45
Southall	12, 24, 26, 32, 97, 100, 102, 103
Southampton	125

Southdown	41, 43, 108
Southend	19, 41, 43, 119, 125
Southend Corporation	18, 80, 128
Speyhawk Land & Estates Ltd	47
Sri Lanka	130
Stagecoach	124, 125, 129
Staines	34, 37, 41, 43, 45
Stamford Brook	27, 32, 61, 66, 103, 106
Stamford Hill	10, 11, 12, 16, 25, 30, 33, 88, 89, 93, 103, 104, 105, 106, 107
Stanwell Buses Ltd	108, 109
Star Rider	113
Stevenage	34, 36, 41, 43
Stockwell	14, 15, 17, 18, 27, 30, 32, 33, 48, 49, 61, 75, 101, 103
Stonebridge	17, 27, 30, 66, 67
Stonegarth	129
Strachan	15
Strathtay	125, 131
Strathclyde	124
Streatham	11, 12, 18, 27, 30, 71, 73, 77, 98, 99, 104, 105
Sunderland	62, 63
Surrey	42
Surrey County Council	26, 108
Sutton	15, 20, 21, 26, 32, 83, 86, 108, 118
Swanley	34, 36, 37, 44, 46, 47
Swift	10, 11, 14, 20, 29, 35, 44, 45, 49, 56, 84, 122
Sykes, Paul	121
Systemtext	118

T

TBP Engineering	121
T&GWU	18
Tate & Lyle	70
Thames Valley	32
Thomson Shepherd	70
Thornton Heath	12, 20, 24, 31, 98, 105
Tinsley Green	44
Titan	19, 23, 25, 27, 31, 32, 33, 41, 43, 56, 59, 61, 63, 78, 98, 101, 102, 113
Tottenham	10, 12, 14, 15, 19, 20, 25, 26, 29, 30, 55, 87, 100, 103
Tourist Bus	103
Tower Hotel	70
Tower of London	101
Trafalgar Square	18
Tram & Trolleybus Department	12
Transmatic	109
Transport, Department of	31
Transport & Road Research Laboratory	11
Transworld Leisure	118
Travelcard	32
Treadmaster	16
Tring	34, 41, 43
Tuhill, David	5
Tunbridge Wells	118
Turnham Green	10, 11, 18, 27, 102, 120
Twickenham	11
Tynesider	63, 64, 85
Tyne & Wear	62

U

USA	16, 121
Underground	118
Underwoods	53
United	62
United Counties	125, 127, 129
Universal Bus Ticketing System	25
Upton Park	11, 14, 19, 20, 21, 24, 25, 27, 32, 55, 91, 94, 101, 111
Uxbridge	20, 27, 32, 105, 120

V

Valley Drive	43
Vernons Pools	52
Verwood Transport	125
Victoria	32, 49, 75, 98, 103, 109, 110, 114, 116, 118
Victoria Line	10, 98
Volvo	61, 98, 106

W

Waltham Cross	103, 106, 113
Walthamstow	14, 15, 19, 21, 24, 82
Walworth	14, 17, 19, 30, 33, 89, 100, 101, 102, 120
Wandle District	30, 108
Wandsworth	11, 27, 30, 105, 119, 123
Warren Street	10
Washington	63
Watford	43
Way, Sir Richard	10
Way & Williams	121
Wearsider	63
Webasto	63
Welwyn	38, 39, 41, 44
Wembley	74, 76
West Germany	123
West Ham	12, 14, 19, 21, 24, 27, 30, 32, 33, 82, 101
West Riding	80
Westbourne Park	30, 32, 77, 96, 106, 111
Western Scottish	131
Westlink	103, 108, 109
Westminster	100
Whatstandwell	129
Whitehall	33
Willesden	17, 19, 33, 72, 78, 103, 111
Wilmslow	131
Windsor	35, 36, 37, 38, 39, 41, 43, 44, 47, 86
Wisdom	53
Wombwell Diesels	44, 46, 64, 82, 83, 84, 121
Wood Green	10, 11, 12, 14, 18, 21, 22, 27, 31, 100, 102, 103, 110
Woolmark	71, 73
Woolworth	70
Wright Signs	38
Wulfrunian, Guy	80

Y

Yellow Pages	50, 52, 53
Yorkshire Traction	129
Youngers	52